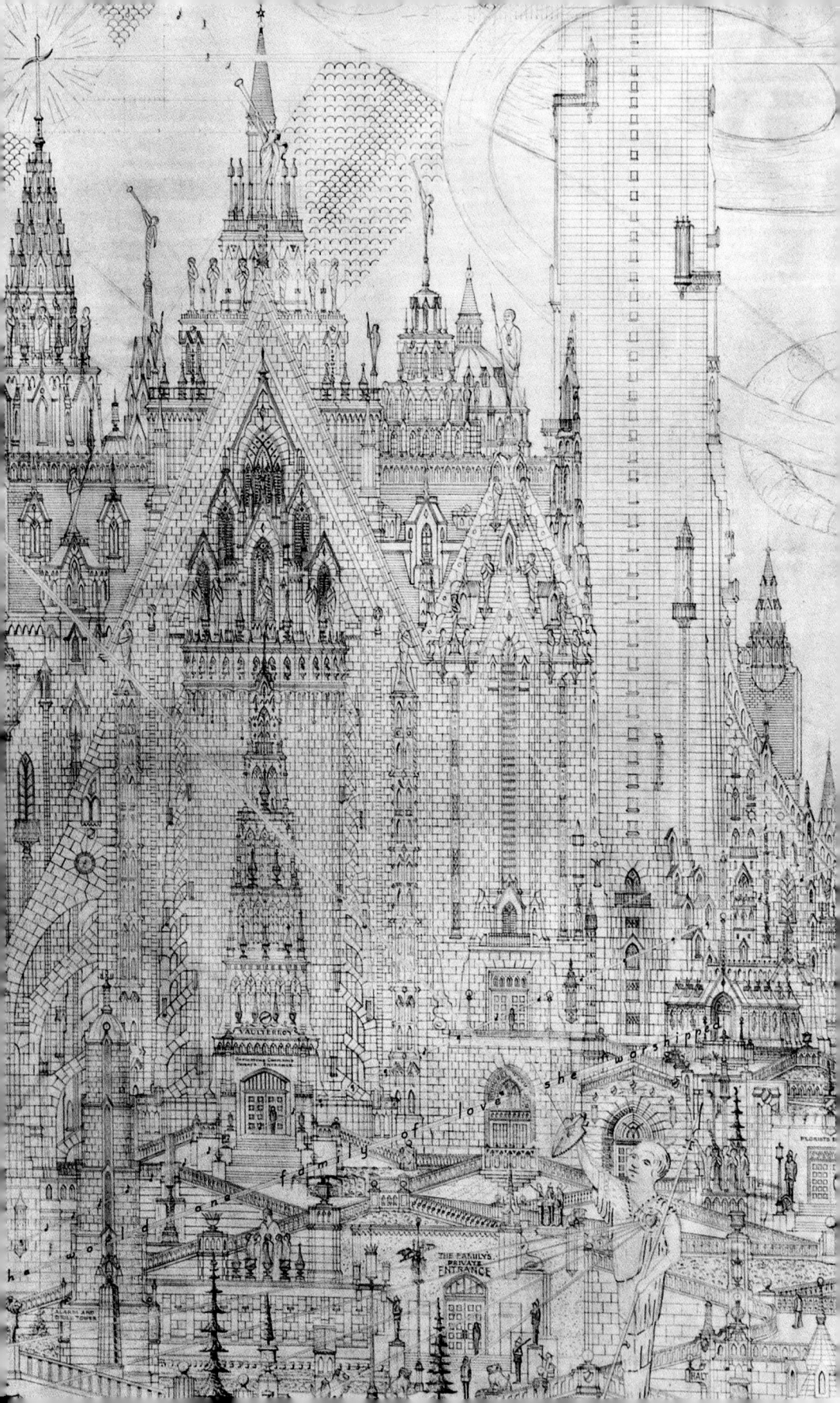

FARFETCHED

Mad Science, Fringe Architecture and Visionary Engineering

Roger Manley, Tom Patterson

Gregg Museum of Art & Design

Raleigh, North Carolina

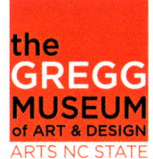

FARFETCHED—Mad Science, Fringe Architecture and Visionary Engineering is published in conjunction with an exhibition of the same title curated by Roger Manley and Tom Patterson for the Gregg Museum of Art & Design, January 17–April 26, 2013. A presentation of ARTS NC STATE.

ISBN 978-0-9831217-2-5

Printed in the USA by Four Colour Print Group

Front cover: *Shock Helmet*, courtesy of Steve Erenberg, radio-guy.net. Phrenology was a 19th-century pseudoscience based on the concept that the brain comprises localized areas that perform different functions. While this is true, practicing phrenologists extended the concept to maintain that the shape of the skull expressed the development of the brain inside. A bump in the area supposedly devoted to "secretiveness" or an under-defined "altruism" zone might indicate a criminal mind; a large forehead might indicate intelligence, etc. Later research showed there is no relationship between head shape and character or intelligence.

This device was made by the Energo Company of Turin, Italy about 1900 and was used for phrenological shock treatments. Electrical connections could be moved to any of the helmet posts to deliver mild, localized shocks in order to cure anything from baldness and headaches to improving benevolence or a sense of humor. [Source: Radio-Guy.net]

Frontispiece: Achilles Rizzoli, *Mother in Metamorphosis Idolized-/The Kathredal (A Glimpse of Heaven's Intermediate Grandeur)*, 1938. 54 x 35 in., graphite, drafting ink. Courtesy of the Ames Gallery, Berkeley, CA.

Back cover: Charles Dellschau, *Airship 44–99*, 1919. 16.75 x 17 in., watercolor, poster paint, pencil, collaged news clipping. Courtesy of Selig D. Sacks.

For our fathers,
Thomas Keys Patterson
and Richard Ernest Manley

Everyone knows what mad scientists are like: wild eyed and even wilder haired (or shaved completely bald), given to fits of maniacal laughter and working alone in their creepy labs except, perhaps, for their groveling subhuman assistant. Down through the ages they've appeared in myriad fictional guises, from the legendary Dr. Faustus to Professor Moriarty, Sherlock Holmes's mathematical nemesis, *Back to the Future*'s Doc Brown, or DC Comics' Lex Luthor, the archenemy of Superman. Even the Wicked Witch of the West in the *Wizard of Oz*, as well as Oz himself, are specimens of the type. Almost always, mad scientists have been portrayed as bent on achieving vast personal power, whether through reincarnating the dead, building sentient robots, inventing super weapons, mastering time, air or space travel, or artificially generating wealth through one questionable means or another.

Their real-life predecessors were medieval alchemists like Heinrich Agrippa and Nicholas Flamel, who sought to change ordinary metals into gold, or the marvelously named physician-astrologer, Philippus Aureolus Theophrastus Bombastus von Hohenheim, who sensibly went by "Paracelsus" for short. Paracelsus was said to have experimented with making an artificial person, a theme revisited by the protagonist in Mary Shelley's *Frankenstein*. Even legitimate scientists like Galileo, Isaac Newton or Nikola Tesla were mocked or labeled "mad" for pursuing astrology and the occult alongside math and science, or because—like Doppler, Darwin, Gauss, Ohm, Pasteur, Goddard, Turing, the Wright brothers, Madame Curie, Barbara McClintock, Hedy Lamarr, or Watson, Crick and Rosalind Franklin—their discoveries challenged long-established scientific or religious "facts" or "laws."

All the ridicule points to age-old anxieties regarding progress and technology. The unease is not wholly groundless. Massive contemporary issues like global warming, population growth, or the threat of nuclear disaster are intricately coupled with the technological advances of the past 200 years. But fear of progress extends much farther back than that, since those who could write, make metal tools, grow crops or split atoms could dominate those who had not yet learned how. Whether the scientists, architects or engineers who develop technology are admired or feared largely depends on whether they are viewed as being on one's own side, in league with competitors, or selfishly on their own.

It's time, then, to remind ourselves that oddball mavericks and outsiders have made many of the greatest contributions to modern civilization. Anyone who flips a light switch, turns on a TV, makes a phone call, listens to a radio, watches a movie, plays recorded music, takes a photo, uses a computer, drives a car, or takes a plane flight has lone eccentrics to thank. All of these were feats once considered mad or impossible—until they were successfully accomplished.

FARFETCHED—Mad Science, Fringe Architecture and Visionary Engineering asks that one keep an open mind, leaving preconceptions behind. Remember that at one time or another belief in germs, quarks, black holes, meteorites, fossils, blood circulation or continental drift was considered laughably wrongheaded. If something presented here makes you wonder whether it is art or technology or something in-between, also try to remember that some of the greatest scientists and inventors were artists as well, and vice-versa. In the end it is *all* creativity, and as such, something to celebrate. As the British mathematician and philosopher Alfred North Whitehead once said, "Every really new idea looks crazy at first."

—**Roger Manley**, Director, Gregg Museum

Bold Explorations in the Fields of Time Travel,
Hybrid Mechanics, Perpetual Motion, Levitation,
Electromagnetic Healing, UFO Technology,
Psychotronics, Radionics, and Visionary Design

TOM PATTERSON

The late Emery Oliver Blagdon, a reclusive mechanic and former hobo, spent the last 30 years of his life alone in a ramshackle house on a Sandhills farm near Platte, Nebraska. Throughout that time he worked daily with bits of plastic, paper, wood, foil and metal, along with wax crayons, bottle caps, soft-drink cans, hairpins, popsicle sticks, glass vials of mineral salts and countless yards of baling wire to painstakingly fabricate more than 400 geometrically complex objects, which he installed in a crudely built shed. He suspended these curious assemblages from the walls and rafters, interspersing them with his paintings—bright geometric designs on wooden panels, sometimes embellished with metal hardware, and typically arranged in stacks on the floor. Interwoven with blinking Christmas-tree lights, the resulting display was visually dazzling, but its primary purpose was therapeutic. Blagdon invariably referred to it as his Healing Machine, indicating his concept of it as a chamber for channeling and concentrating the earth's electromagnetic energies to alleviate pain and cure illnesses. In the years prior to Blagdon's death at 78 in 1986, local

Emery Blagdon, components of the *Healing Machine*, largest 79 x 14 x 7 in., wire, copper, foil, tape, lids, found objects. Courtesy of Cavin-Morris Gallery, Selig D. Sacks, Mark and Kathryn LeBaron, and Karen and Robert Duncan.

residents suffering from arthritis and other ailments came to sit in the Healing Machine and absorb its concentrated energy. Some visitors reported being able to feel the machine's electrical charge from the moment they stepped inside the shed.[1]

Since the late 1960s Paul Laffoley, a Harvard-educated architect and prodigious polymath, has worked in a 540-square-foot room he calls the Boston Visionary Cell, where he makes in-

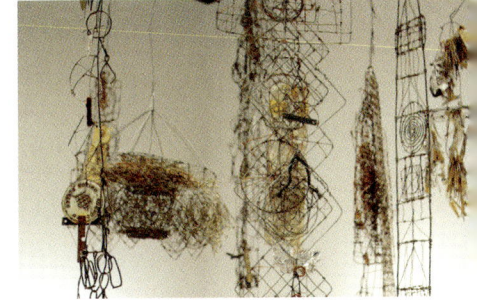

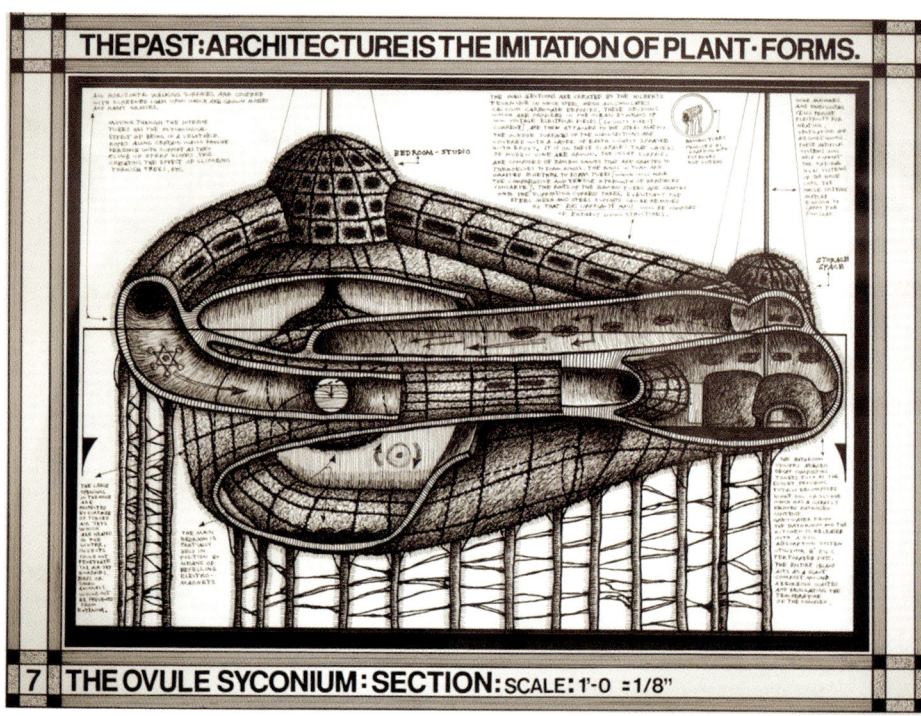

THE PAST: ARCHITECTURE IS THE IMITATION OF PLANT·FORMS.

7 | THE OVULE SYCONIUM : SECTION : SCALE: 1'-0 =1/8"

Paul Laffoley, detail of *Das Urpflanze Haus II* (5–8), 1993–95, 27 x 36 in., drafting ink and presstype. Courtesy of the Cartin Collection.

tricately structured paintings, diagrams, drawings and three-dimensional models based on his uniquely innovative design concepts and his investigations into areas such as time travel, perpetual motion and psychotronics. A characteristically multifaceted example of Laffoley's projects is *Das Urpflanze Haus* (the "Primordial Plant House"), his design for a labyrinthine, multi-tiered "living building" largely covered with and structurally inseparable from a network of plants and trees whose branches are grafted together to form a single plant with a multiple root system. Evidence uncovered by recent explorations of Mars has led him to propose that our nearest neighbor planet might be an ideal environment for this botanically enlivened structure. Laffoley has suggested that such designs and some of his other ideas might be generated or influenced by a tiny metal chip that has appeared in X-ray scans of his brain. He suspects the mysterious chip might be a sophisticated nanotechnological device implanted by extraterrestrials.[2]

Working in old warehouses in Asheville, North Carolina, over the last ten years, Sean Pace (aka "Jinx") has spent long hours cannibalizing salvaged machinery and ingeniously combining spare parts to build improbable, post-industrial kinetic devices. His outlandishly deconstructed and re-engineered contraptions serve as humorous critiques of consumer culture and contemporary social trends. His most ambitious work to date is his ensemble *Fight or Flight*, which consists of two separate components. To build the aggressor component, known as the "Chicken Shooter," he cobbled together a dentist's chair, the rear axle from a pickup truck, a set of industrial wheels salvaged from a defunct furniture factory and a 20-year-old Honda motorcycle with its front end cut off and turned around backwards, among other parts and artifacts. The resultant hybrid machine suggests a piece of imaginatively improvised, scrapwork artillery for a post-apocalyptic battlefield—a missile-firing cannon operated from an open-air gunner's seat.

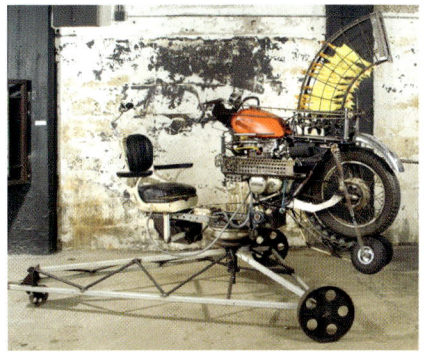

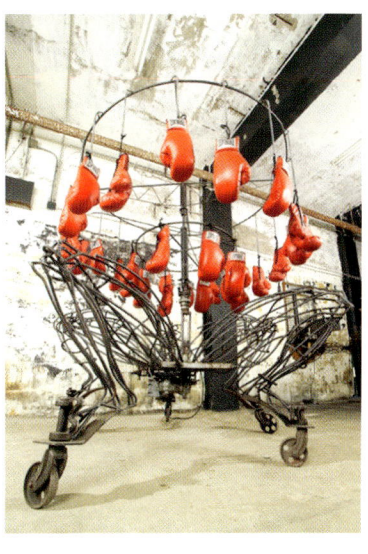

In fact that's exactly what it is, except that the missiles it's designed to fire are rubber replicas of dead, plucked chickens, of the kind traditionally sold in novelty stores. The "Shooter" is designed to fire these harmless, floppy projectiles across up to 50 yards of open space toward its intended target, a combatively defensive gizmo known as "The Boxer." This towering carousel of 28 hanging boxing gloves, animated by an old washing-machine motor, spins back and forth, causing the shiny red-leather gloves to alternately swing outward and jerk back, knocking the rubber chickens from the air as they're fired toward it. Set in motion together, the two devices perform their own outrageous *Ballet Mechanique*, which Pace has conceived as an automated, slapstick comment on overpopulation, domestic violence and society's inability to address urgent global problems.[3]

Sean Pace, *Chicken Shooter* (at top), 2006, 90 x 102 x 156 in., and *The Boxer*, 2006, 90 x 84 x 84 in., mixed media. Courtesy of the artist.

Although they're from very different backgrounds and motivated by varying interests and concerns, Blagdon, Laffoley, Pace and the other individuals represented in **FARFETCHED—Mad Science, Fringe Architecture and Visionary Engineering** share several key characteristics relevant to this exhibition: They've all set out to solve problems or answer questions in the areas of science, design and/or engineering, but they've gone about it in ways that professional specialists would likely consider improbable, unconventional, eccentric, outrageous or even insane. Fortunately, skepticism on the part of experts and academics has failed to deter these brave, deeply committed souls as they've forged ahead in their often lonely pur-

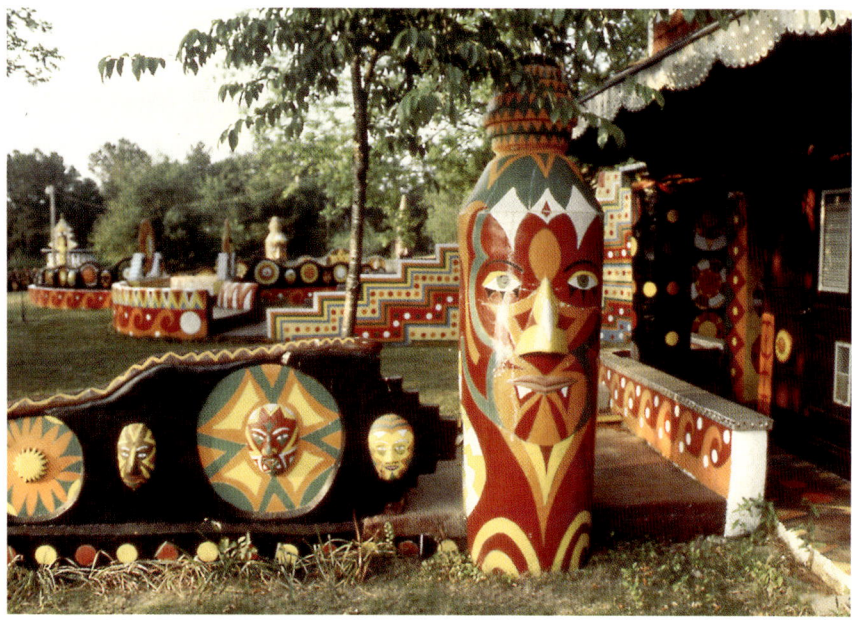

4

suits. On the whole and to their great credit, they've demonstrated scant concern for public opinion or professional conventions.

The intense energy and long hours these inventors, artists, designers and alternative scientists give to their unusual projects and experiments seems to derive from an intuitive sense that they're "onto something"— the proverbial light-bulb glowing over the head—and an urgent compulsion to follow it through at all costs. Some of them have achieved breakthroughs, some have failed to accomplish their original goals and others have undertaken projects that could theoretically be pursued endlessly without resolution. In all cases, though, these creative dreamers and visionaries have managed to produce tangible results—working devices, design drawings, rough sketches, paintings, scale models and other objects—that are intriguing, often strangely beautiful and sometimes astonishing in their conception and design, if not always in their operation.

Eddie Owens Martin, an artist and self-proclaimed psychic who called himself "Saint EOM," had visions of exotic-looking people wearing special suits that enabled them to levitate and travel through the air above the earth's surface. Instead of dismissing these as impossible fantasies, he took them seriously, creating countless drawings and numerous paintings of people wearing his levitation suits. He also depicted them in sculptures integrated among the sinuous walls, columnar totems and customized pagodas of his visionary masterwork, the strikingly painted four-acre environment known as Pasaquan, near Buena Vista, Georgia. Martin described his vision of such a uniquely utilitarian garments as "a complete elastic suit that puts pressure on the right parts of the body, so that man can control levitation of his body and be lifted up and keep himself away from the pull of gravity, so he can act normally out yonder in space. Man has that power in his body and don't know it. Them suits haven't been developed yet, but I've seen 'em in visions that the spirit has shown to me. They'll be pressurized, and they're all drawn together with elastic bands, and there's a frame thing that comes in here and covers up part of his face, and there's glass over his eyes so he can see out of it. It'll be air-conditioned and everything."[4]

Attempting to create a prototype, Martin was able to fabricate something that visually approximated what he'd seen—an elasticized suit of sewn-leather discs and bands, carefully measured to fit his body. But he never managed to figure out how to efficiently air-condition it or attain liftoff while wearing it. Although the experiment was unsuccessful from the standpoint of his original vision, the levitation-suit design was integral to his hybrid esthetic, yielding a number of striking, idiosyncratic images, including some of Martin's best paintings and sculptures.

As different as they were in appearance and philosophy, Martin had a lot in common with his friend Howard Finster, who lived 200 miles to the north in Pennville, Georgia. A preacher and all-purpose repairman, Finster transformed his backyard into a sprawling art environment that attracted visitors from all over the world, and he made drawings and paintings based on visionary experiences he could describe in elaborate detail. Finster's multi-acre environment, popularly known as "Paradise Garden,"

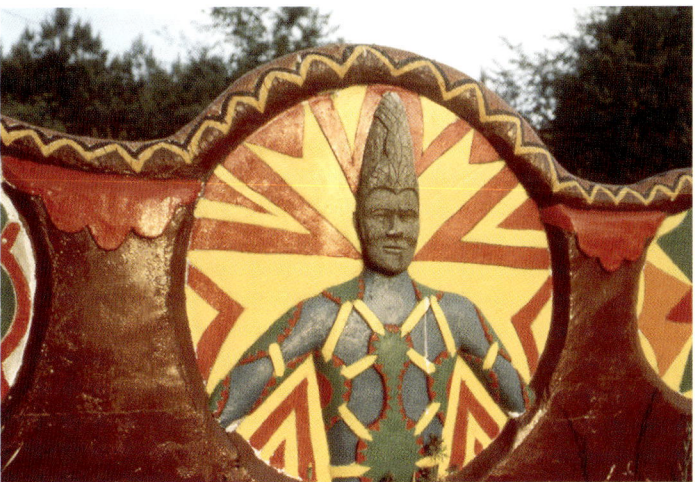

Eddie Owens Martin, wall with figure in levitation suit and up-swept hair at *Pasaquan.*

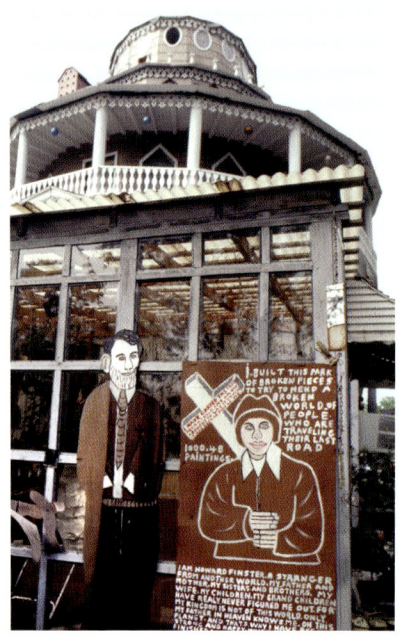

6

incorporated fanciful buildings, concrete sculptures, outdoor paintings and an extensive if haphazard collection of the "Inventions of Mankind."

Finster also had an abiding interest in perpetual motion. In addition to drawings and paintings of devices he called perpetual-motion machines, he drew other unusual mechanisms and unconventional vehicles he claimed to have seen in visions of "other worlds beyond the light of the sun." The largest structure in his garden, which he dubbed the World's Folk Art Church, is an unwieldy hybrid structure surmounted by a silver-steepled, four-story wedding-cake tower built around a spiral staircase. It represents his attempt to build one of the idiosyncratic, multi-tiered "heavenly mansions" he also attributed to his visions and depicted in literally thousands of paintings and drawings.[5]

Howard Finster, *World's Folk Art Church*, Pennville, Georgia, built in late 1970s. Photo: Roger Manley, 1983.

Otherworldly architecture glimpsed in a visionary state also provided the inspiration for Richard Brown, an independent florist in Littleton, North Carolina. Unlike those of Eddie Martin and Howard Finster, Brown's visions had an apocalyptic, dystopian character. What he saw were the rusted, towering structures of a militaristic future civilization that had collapsed—a "future past," as he characterized it. Furthermore, this panorama of industrial and technological ruin was also observed by the mysterious occupants of glittering spacecraft and oddly configured helicopters that swooped and hovered overhead. Brown was so captivated that he was compelled to reproduce it all on a miniature scale in the back room of his otherwise ordinary florist's shop, using nothing but the supplies and materials of his trade. He began this curious venture in the mid-1990s, while undergoing an extended period of personal difficulty.

To create his imaginary cityscape of skyscrapers, cranes, derricks, aircraft carriers, warships and structures resembling offshore oil rigs, he carved pieces of the floral foam he typically used to make wreaths and funerary displays, then embellished them with corsage pins and floral wire, using hot glue to attach and affix some of these materials. He coated the rough foam surfaces of the terrestrial structures with olive-drab spray paint to lend them a patina of age and then distributed a number of matchbox-size commercial models of tanks and other military and industrial vehicles on their decks, as if they'd been left there by this world's vanished inhabitants.

He employed the same materials and techniques to create the spaceships and other aircraft patrolling above these structures, but more often used black or green spray paint on their surfaces, which he also decorated with glitter, presumably to distinguish them from their ostensibly deteriorating earthbound counterparts. Brown used filament wire to suspend these airborne vehicles from the ceiling. The entire installation suggests a scale-model set for a low-budget Japanese sci-fi movie, with a monster dragon about to enter and trample everything to smoking wreckage while batting away the UFOs.[6]

Ionel Talpazan claims to have actually seen a UFO, in a real-life "close encounter" he vividly recalls from his troubled childhood in rural Romania. His twin brother died shortly before his own premature birth, and his unmarried parents abandoned him when he was six. Soon afterward he was adopted, but his alcoholic foster mother beat him frequently. One night when he was seven, he fled the house during one of her violent outbursts and hid in a ditch. There he soon found himself surrounded by blue light radiating from a huge disk hovering silently in the sky above him. This life-changing experience is the source for the hundreds of drawings and paintings he has since made depicting multicolored, disc-shaped aircraft.

Romania was still controlled by a repressive, Soviet-style Communist

Richard Brown, spaceships from *The Future Past*, florist's foam, wire, glitter, faux pearl pins.

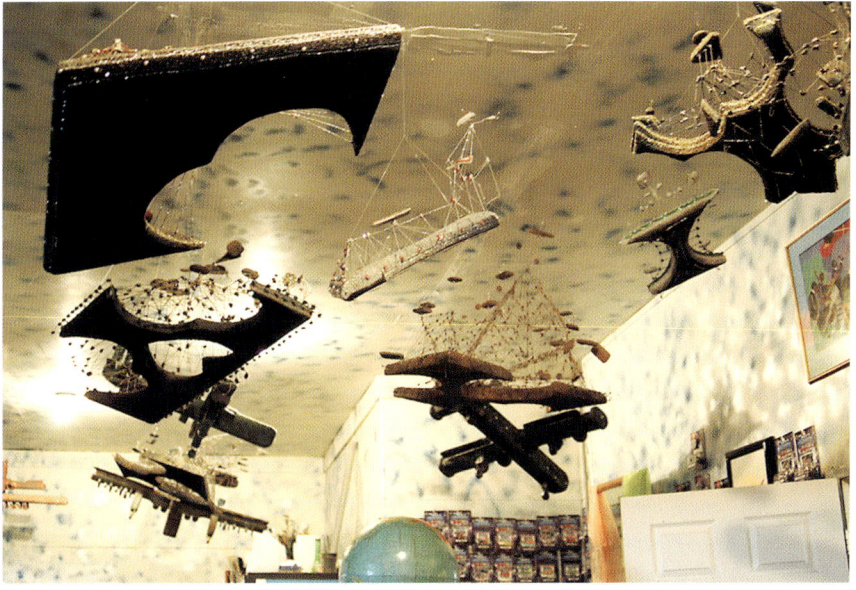

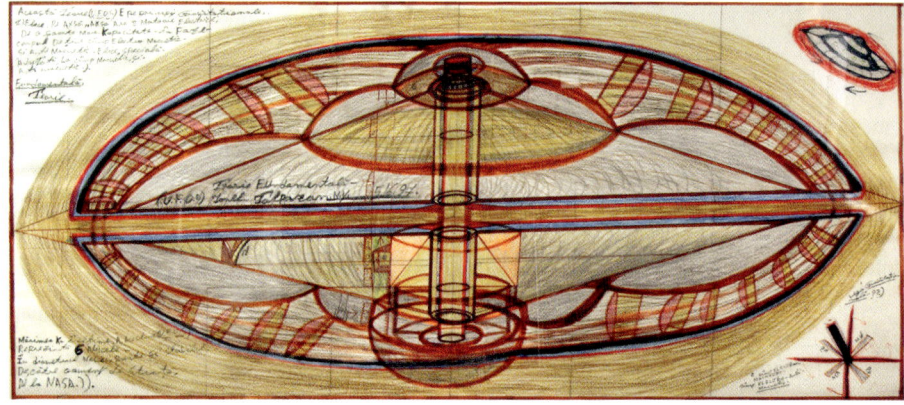

8

government in 1987, when Talpazan, then in his early thirties, fled the country. After the U.S. government designated him a political refugee he made his way to New York, where he worked occasionally as an unskilled laborer and soon returned to his practice of drawing UFOs. By 1990 he was regularly selling his drawings on the Manhattan sidewalks.

In most of his drawings Talpazan depicts spaceships in cross-sectional diagrams that appear to reveal something of their internal workings, and he often incorporates hand-lettered, explanatory texts in Romanian or phonetically spelled English. He has also made a number of sculptural models of UFOs, including a large, painted plaster model occupying much of the space in his small apartment in Harlem. In his more elaborate, mixed-media drawings the colors are coded to indicate various energy sources he believes are used to power these flying craft, including nuclear, solar and electromagnetic. He believes NASA has much to learn from his drawings and his theories on UFO technology, including his ideas about advanced propulsion systems and accelerated space travel.

Talpazan exhibits an almost evangelistic fervor when talking about his UFO encounter, his theories about UFOs and the beliefs he has developed about them: "My art shows spiritual technology, something beautiful beyond human imagination, that comes from another galaxy. Something superior in intelligence and technology. So, in a relative way, this is like the God, it is perfect . . . . My art is about the big mystery in life. How did we get here on planet Earth? Why are we here? Is there life on other planets?"[7]

George Widener suspects that the secrets of life might be contained in numerical systems and sequences. He has had a lifelong obsession with time, numbers and calendars, as well as a prodigious ability for mental calculation and memorizing dates. When he was in his early twenties he was an aerial-surveillance technician for the Air Force and studied engineering, but he subsequently spent more than a decade homeless during a period of intense social withdrawal that left him ill-equipped to handle a steady job. He slept in parks and abandoned buildings in a string of different cities in the U.S. and Europe, and otherwise spent much of his time filling notebooks and scrap paper with numbers, calendar grids, facts, statistics and drawings. In the late 1990s counseling enabled him to understand and overcome some of his problems.

Around that time he began combining his numerical interests and drawing skills to create increasingly elaborate images and diagrams intended to illustrate his ideas and some of the insights to which his calculations have led him. These works "explore symmetrical ideas that take place in our brains," in his words, and relate to his conviction that all of us have potential access to a vast, subconscious memory bank. Images of architectural settings, transport vehicles and imaginary devices for computation or time travel are central to most of his recent works. The images are typically set off against dense "fields" composed of sequential calendar dates in tightly spaced lines, while many of these dates have special connections to one another, Widener explains. The hidden codes, ciphers and interconnections among these dates are undetectable to most viewers of Widener's work—and even to himself in some cases, he is certain—but he says they will be easily read by "intelligent machines/cyber people of the future." He predicts that this future audience will be able to scan, instantly recognize and derive something akin to aesthetic pleasure from these numerical patterns and correspondences: "Robots will one day love and collect my work," he insists.

Projecting even further into the future, Widener suggests, "As we leave

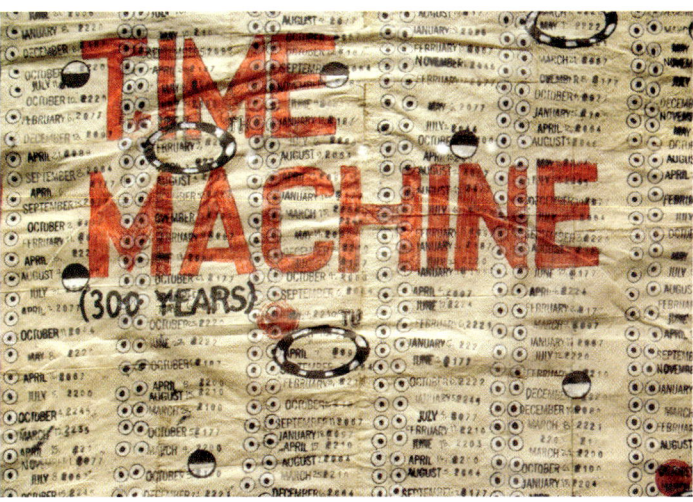

George Widener, detail of *Time Machine*, 2011, 63 x 23.5 in. Rubber stamp ink, watercolor on paper towels. Courtesy of Henry Boxer Gallery, London.

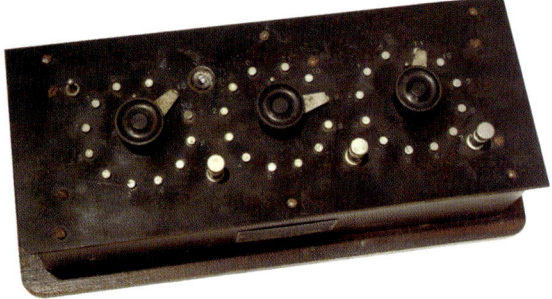

The Abrams Oscilloclast

Dr. Albert Abrams (1863–1924), famed discoverer of the pulmonic and cardiac reflex, is generally considered the father of radionics. Director of Clinical Medicine at Stanford University, fellow of the American Medical Association (AMA) and author of numerous books, in 1919 Abrams and inventor Samuel Hoffman teamed up to create the *Oscilloclast*. Its purpose was curative. It was intended to subject the patient to a negative electrical charge with radio-frequency electromagnetic pulses in-between. In use, the resistance box, including the prominent dials, allowed for different settings or "rates," and was placed between the *Oscilloclast* and an electrode placed upon the forehead of the patient. [From the collection of Duncan Laurie]

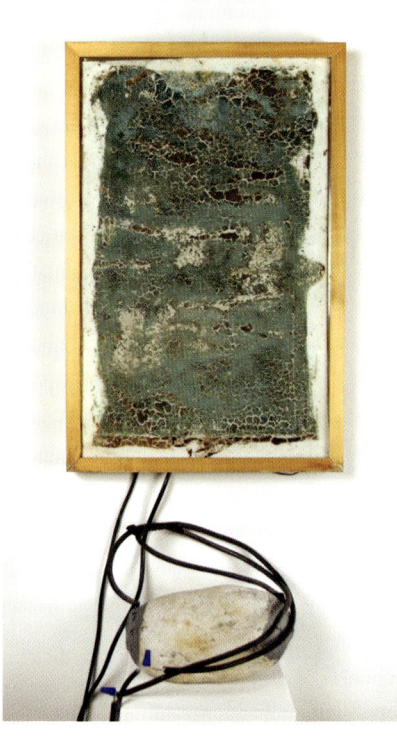

Glass Bio-Sensor

This sensor is made in the style of Orgone accumulator blankets, with alternating metal and organic layers of felt and wire mesh screen. The felt has been saturated with liquified fungal matter to provide a biological signal source, which the Orgone accumulator amplifies. The signals nest within the electrostatic millivoltage output produced by the device. Those voltage gradients are sent to a rate-of-change converter, where the mix can be sonically heard as random pitch shifts when amplified. [Duncan Laurie, *Bio-Sensor*, ca. 2004, 20.5 x 13 in., stone 5 x 8.5 x 4.5 in.]

this planet as a species and explore other systems (with other calendar systems), perhaps several centuries from now my pictures will be a sort of artifact of our own solar system here that will probably be more realized as time goes on."

"I like to think I'm exploring something that's eternal," he has said, adding that he wonders whether his work might eventually yield a new understanding of time and the unfolding of human events.[8]

New insights into the nature of reality and human experience can arise from any number of circumstances. For artist and designer Duncan Laurie it was an encounter with a culture very different from his own that sparked a change in his worldview. In the early 1980s, while visiting the Hopi Indian Reservation in Arizona, Laurie observed a traditional rain dance followed within half an hour by a torrential thunderstorm. This struck him as remarkable in a high-altitude desert where rainfall averages only six to eight inches a year. He had read about shamanic rituals, but witnessing a ceremony that appeared to deliver such immediate, decisive results prompted him to begin rethinking his earlier assumptions about art and its relationship to the natural world. From his perspective as a contemporary Anglo-American, the rain dance was a performance—a work of art. Why, he wondered, had the academic art world ignored or overlooked any potential that art might have to affect natural forces? This question stimulated an ongoing inquiry that soon led him to investigate psychotronics (the study of mind-body-environment interactions) and radionic technology, invented and developed during the early 20th century for use in healing. He began an intensive study of radionics, bought the first in what would become a growing collection of radionic devices and started conducting independent research and creative activity in the field.

As an artist, Laurie works in two- and three-dimensional mediums ranging from glass to printmaking, and as a professional designer he specializes in architectural glass. In keeping with his longstanding interest in indigenous cultures, he has closely examined petroglyphs at several sites in the southwestern and northeastern United States. His studies have convinced him that these glyphs and other ancient rock art were made to communicate with and have specific effects on nature, like the Hopi rain

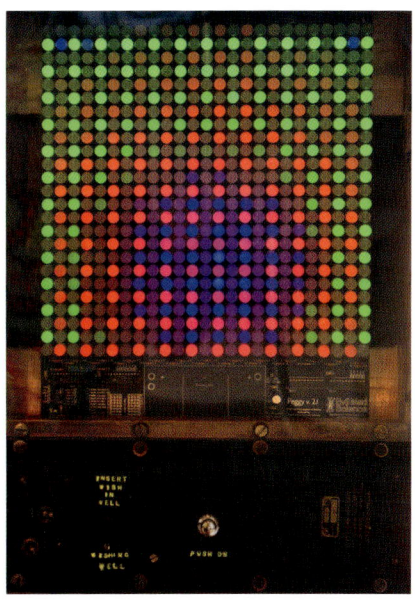

Duncan Laurie, detail of LED display from the *Purr Generator*, 2011 (see page 30).

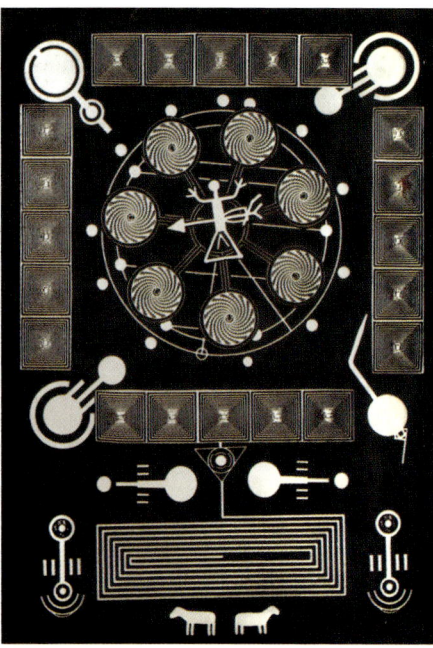

Radionic Circuit With Sheep

This playful cyberglyph combines functionality with metaphor. Drawing upon crop circles, magical icons, and psychotronic geometries, the glyph pays homage to the dot-matrix radionic designs of Darrell Butcher which date to the 1950s. Though Butcher was an aircraft engineer, he couldn't operate existing radionic technology, so he created his own extremely artistic technology that was altogether free of electronics. For operational energy, Butcher accessed what he called a "downpouring" of vital energy to treat patients, simply through touching the device. His technology drew upon aesthetics: Archimedean spirals, pegboard contraptions, graphic 'rate' cards and other purportedly scientific representations. Butcher's devices are still considered among the most imaginative of all radionics works. [Duncan Laurie, 1993, 13 x 9.5 in., printed circuit.]

dance. For that reason he now views them as precursors and primal analogs of radionic devices.

In his book, *The Secret Art: A Brief History of Radionic Technology for the Creative Individual*, Laurie summarizes the largely forgotten history of radionics from its discovery by Albert Abrams, a prominent neurologist, through its various permutations and applications in this country and elsewhere over the last 100 years. He also takes a critical look at its rejection by the scientific and medical establishments after decades of promising research and successful treatments. He makes an eloquent case for the enduring validity of radionics in and beyond its medical applications. "I look at it more as a metaphor for how human intent can be employed to transform energy and perform work," he writes. "In the simplest sense, radionics is a methodology by which Information can be used to move Energy . . . ." Its history "clearly shows a progression from a mechanical and electrically based technology towards one of aesthetics, information and ultimately spirit. As radionics has become less and less comprehensible in conventional scientific terms, it has become more accessible in artistic terms."[9]

Laurie pursues his research into radionics, subtle energy and nature intelligence in a radionically designed, three-story building in Jamestown, Rhode Island, with all-around window walls and glass-brick floors. In that appropriate setting he maintains an ongoing collaboration with electrical engineer Gordon Salisbury (and, from a distance, colleague Todd Thille) to develop and use new forms of radionic technology, some of which involve sonic communication with plants, rocks or other components of the natural world.[10] Their artistic collaborations have involved devices such as radionic socks (imprinted with images of radionic circuitry designed to aid in wish fulfillment) and a more complex, functionally related device called the *Purr Generator*.

Modified from an earlier radionics device called a BETAR (an acronym for 'Bio-energetic transduction-aided resonance'), the Purr Generator represents a collaborative project on which Laurie worked with Salisbury, Thille and Michael Bradford. It consists of a narrow bed incorporated into a cuboctahedron frame (a Buckminster Fuller dymaxion form), augmented with special LED lighting and a sound system that electronically amplifies the sustained purr of a cat, intended to facilitate relaxation and healing.

Laurie contributed the generator's radionic component, the "input well," into which the person lying on the bed inserts a wish or request written on a piece of paper. This small opening contains a coil hardwired into the electronics system, thereby radionically linking the user's request to his or her experience in the device. It's a contemporary variation on the archetypal wishing well, about which Laurie notes, "The universal enjoyment of wishing wells of all types and sizes is ample testimony to their effectiveness."[11]

As an integral part of his research, Laurie creates drawings, prints and two-dimensional mixed-media art whose imagery he appropriates from earlier radionic devices and related sources. Some of these works are actually designed to function radionically, like contemporary equivalents of petroglyphs, while others combine functional and metaphorical components. Laurie has suggested that ancient rock art functions by modifying and directing telluric currents within the landscape, and in that respect is comparable to the geometric patterns on electronic circuit boards, which perform essentially the same function with electrical currents. This comparison is the basis for his "Cyberglyphs" series, made as printed circuit boards. Some of these works incorporate images borrowed from the giant, geometric-abstract-patterned "crop circles" that have mysteriously appeared in grain fields in several countries, mainly in southern England, since the 1970s. Laurie has theorized that these monumental designs are closely akin to petroglyphs, in that they're expressions of some as-yet-undeciphered intent inscribed directly on the earth by unknown creators.[12]

Among its other implications, Laurie's research highlights the role the mind plays in defining our experiential limits. Our expectations and beliefs have everything to do with how we apprehend the world and navigate our way through it. They can either leave us open to new possibilities and novel solutions, or blind us to such options. To dismiss an unconventional proposition as impossible, unscientific, superstitious, absurd or crazy is to risk missing an opportunity.

A skeptical intellect is fundamental to rigorous critical thinking, a skill too rarely exercised in our society. But a closed mind is the enemy of art, invention, experimentation and everything else this exhibition extols. *Farfetched* celebrates the open, intuitive, visionary mind in its application to science, building design and engineering. The 38 artists, inventors, researchers and designers represented in this exhibit have all made intriguing and extraordinary contributions. Their ideas and the things they've made, designed and/or built might seem "farfetched," but they're also heroically aspirational and inspirational. Together, they make a profound statement about the interworkings of intelligence, intuition and creative vision—about the human mind continually pushing beyond pre-established limits in order to keep expanding our options.

NOTES

1. Most of the information on Blagdon is from Leslie Umberger, *Sublime Spaces & Visionary Worlds: Built Environments of Vernacular Artists* (New York & Sheboygan: Princeton Architectural Press and John Michael Kohler Arts Center, 2007), pp. 204–223.

2. Information on Paul Laffoley is from the author's phone interviews with the artist in the summer and fall of 2012, and from "Paul Laffoley" in Roger Manley, et al., *The End is Near: Visions of Apocalypse, Oblivion and Utopia* (Los Angeles: Dilettante Press, 1997), pp. 74–80.

3. Information on Pace is from the author's discussions with the artist in the fall of 2012 and from the artist's website, www.seanpace.com

4. As quoted in Tom Patterson, *St. EOM in the Land of Pasaquan* (Winston-Salem: Jargon Society, 1987), p. 216.

5. See Howard Finster and Tom Patterson, *Howard Finster, Stranger from Another World* (New York: Abbeville Press, 1989) and J.F. Turner, *Howard Finster: Man of Visions* (New York: Alfred A. Knopf, 1989).

6. Information on Richard Brown and his work comes from the author's discussions with Brown during visits to his studio in 2001 and more recently in the fall of 2012.

7. As quoted in Daniel Wojcik, "Mysterious Technology: Daniel Wojcik encounters the flying saucers and visionary art of Ionel Talpazan," in *Raw Vision* #45, Winter 2003–04, p. 56. Some of the other information on Talpazan is also from this source.

8. Quotations from Widener not otherwise attributed are from the author's interviews with the artist in Asheville and Madison County, North Carolina, in June 2010 and via e-mail more recently, in the fall of 2012. Widener's ideas about super-

human intelligence are inspired in part by his reading of Ray Kurzweil's theories as detailed in his book, *The Singularity is Near: When Humans Transcend Biology* (New York: Viking, 2005).

9. Duncan Laurie, *The Secret Art: A Brief History of Radionic Technology for the Creative Individual* (Anomalist Books, 2009), pp. 11, 82.

10. Other collaborators who have worked with Laurie in his nature/sonic experiments include Steve Nalepa, Richard Devine, Justin Boreta, David Last and Aerostaticmusic. In recent e-mail correspondence (12/18/12) Laurie writes, "Most electronic musicians have their computers and themselves to work with in any given composition, unless they are collaborating with other musicians. With the capacity of nature signals to spontaneously respond (and evolve responses) to the closed feedback loop between the composer and his instrument, a seriously new creative ingredient has become available to the electronic musician—direct collaboration with Nature at a primal source."

11. Quoted from an e-mail sent by Laurie, 12/18/12.

12. For more information on Laurie's research see www.duncanlaurie.com.

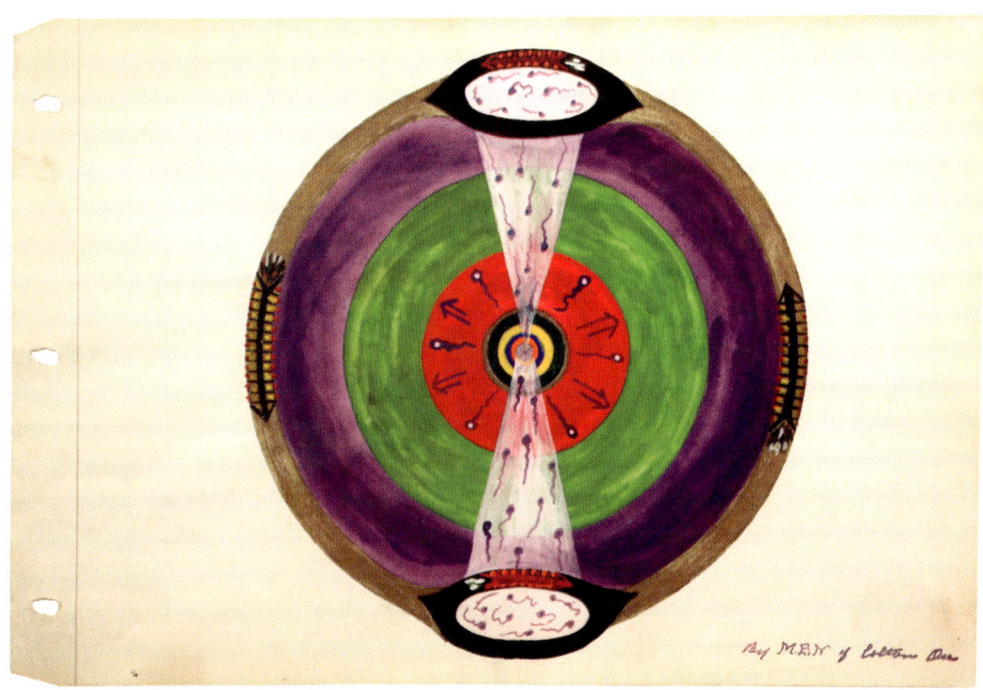

Melvin Edward Nelson,
Futurist Vision, ca.
1965, 11.5 x 16.5 in.,
pencil, poster tempera,
watercolor. Courtesy of
Cavin-Morris Gallery,
New York.

IT'S NOT A BELL CURVE
WITHOUT THE KNIFE EDGE

ROGER MANLEY

My earliest recollection of NC State dates back to a chilly day in 1961 while my family was stationed at Seymour Johnson Air Force Base in Goldsboro. I was in third grade and still eight months shy of my tenth birthday when one morning at breakfast my dad said, "How would you like to stay home from school today and ride with me up to Raleigh? I've got something I need to take and get tested, and you can come along if you want." Naturally, I jumped at the chance. Both of us dressed in civilian clothes, climbed into the family car and set out through the flat pig farm and tobacco country that separates Goldsboro from the state capital.

When we arrived at State College—it wouldn't be renamed NC State University for two more years—we parked near a brick building identified as the chemistry department. Dad opened the trunk and took out a heavy, shoe-box-sized wooden box and carried it in the building where some people were already expecting us. After shaking our hands, a lab-coated technician led the way back to a workroom where he opened the box and lifted out a hunk of twisted gray metal and placed it on a shiny black table.

At first the man in the white coat acted as if this kind of thing was routine, and no doubt it was; all kinds of people were constantly bringing in things to be tested, from soil samples and blown radio tubes to wilted tobacco leaves or failed rivets. He picked up a pair of heavy pincers, nipped off a fingertip-sized piece of the metal and inserted it into the refrigerator-sized machine behind him.

As he got ready to fire up the device, he casually turned and said, "So, mister, what are we looking for?" My surprise at hearing the civilian term "mister" applied to my dad instead of his military rank was immediately capped by his dead-serious response. "Never mind what we're looking for," he said, "you just tell me what you find." The chastened technician quickly lost his smile and turned back to concentrate on operating the machine, which had already begun to hum.

Suddenly an intense light, bright as an arc welder's flash, outlined the edges of the panel that he had closed after placing the sample inside. As he adjusted the dials, a narrow paper tape scribbled with wiggly bands and markings began spewing from a slot in the side of the machine. The lab

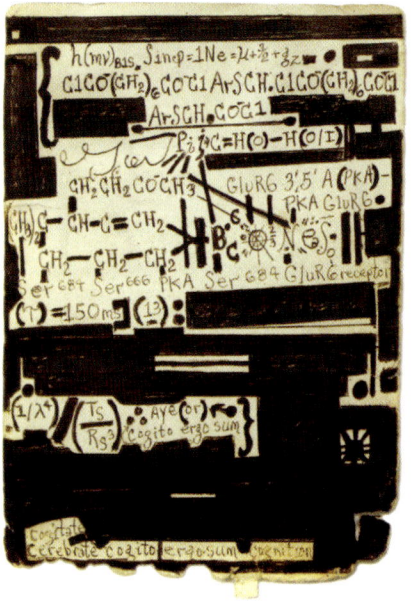

Melvin Way, *Cogito Ergo Sum*, n.d., 5 x 3 in., black ink. Courtesy of the Gallery at HAI, New York.

worker lifted the tape and held it closer to study it. His eyebrows raised with a look of surprised recognition and he muttered something like, "Oh my gosh." As soon as he said that, my dad reached over and snatched the paper strip from his hands and tore the rest of it from the machine. "Let's not talk about this," he said, "and I think it would be a very good idea on your part if you didn't go around mentioning it to any of your colleagues, either." Then he told the man in the lab coat he would need to gather up every last grain of whatever remained from the tested sample and put it back in the box with the rest of the dirty metal glob. Looking a little rattled by now, the man gulped and nodded as he silently and gingerly wrapped everything tight again. Then my father methodically re-sealed the box and motioned for me to follow him out to the car.

My eyes must have been big as saucers, but he only focused intently on finding his way back to the Goldsboro highway. Once we were out in the country again, I got up the nerve to ask him what had just happened. "I'm just trying to solve a little problem," was all he would say. "I'll tell you about it sometime, but I can't tell you right now." Turning the encounter into a mystery, of course, had the effect of burning that whole afternoon so vividly in my memory that I would never forget it.

Years later I found out what had happened. Weeks before, a B-52 had crashed north of Goldsboro with two thermonuclear bombs on board. While one of the bombs had been relocated and secured within hours, the other had slammed into a muddy field near the crash site with such tre-mendous force that it had been embedded deep in the ground. My dad, whose military specialty had to do with the care of nuclear weapons (as I also learned later), was in charge of the recovery. With as much secrecy as possible, his team dug for weeks, occasionally discovering bits and pieces of the H-bomb that had been torn off along the way as it had penetrated deeper and deeper into the earth. The glob he'd taken to Raleigh was one of these pieces. Although it and several other parts were eventually recov-ered, the effort to reach the main part of the bomb eventually had to be abandoned, because groundwater had flooded the site so badly they could no longer proceed.

Today, the bulky uranium fusion chamber that would have made this

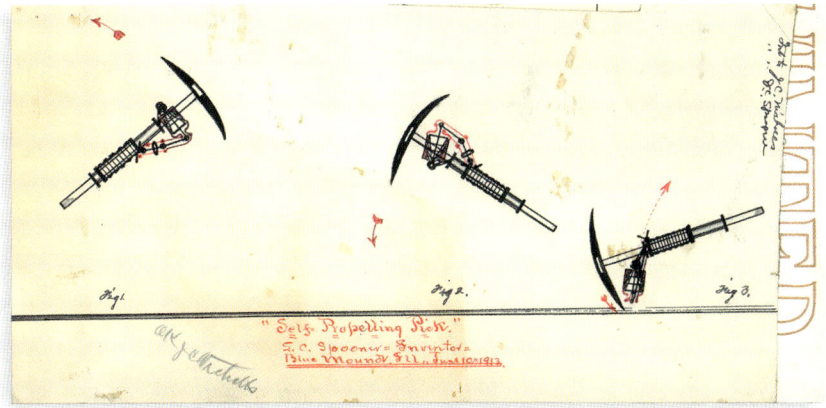

Lee C. Spooner,
*Self-Propelling
Pick*, 1913, 4.5 x
9 in., graphite
and ink. Courtesy
of The Museum
of Everything,
London.

bomb hundreds of times more powerful than the ones that flattened Hi-roshima and Nagasaki still remains buried some 180 feet beneath an oth-erwise non-descript eastern Carolina field. Based on what he had learned during that trip to Raleigh, my dad knew it was no longer in any danger of exploding, and once the soil in the refilled pit had settled, it was deemed safe enough to farm the land again, as long as no one ever went drilling for gas or water nearby. My dad was willing to stake his life on this, and in fact he and my mom eventually returned to Goldsboro to retire within only a few minutes' drive of the "lost" nuclear warhead.

THAT VISIT TO THE NC STATE LAB was not only my first impression of the campus where I now work, but was also, arguably, my first real experi-ence of "mad science."[1] Strange as it was, though, that battered chunk of H-bomb was far from the strangest thing that anyone has ever brought to the university. Here at the Gregg Museum, for example, we get presented with oddities to examine or consider on a near-weekly basis, from ancient mummy wrappings or angels made of driftwood, to things like miniature circuses, duct-tape dresses, and even—only a few weeks ago—Hitler's mirror. Along with Chinese imperial robes and African shaman's costumes we've been offered magical bulletproof undershirts worn by Thai gang-sters, religious mud sculptures, corn art, items made from construction scraps or lizard skins, fishnet stockings made of pink glass, and textiles made with fish scales and beetle wings. Some are gladly accepted into the

1. In this case the term is ironically literal. The Cold War doctrine of military strategy and national security policy in place in the early 1960s was called Mutually Assured Destruc-tion (MAD), in which both the US and Soviet Union promised each other to respond to any attack with a full-scale nuclear retaliation so devastating that neither side could possibly "win."

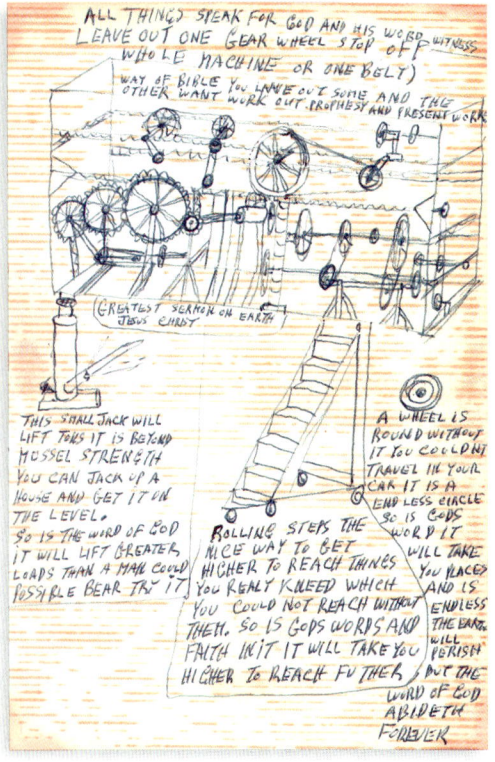

collection, while others—like the afore-mentioned German dictator's mirror—are kindly declined.

And we're just the art museum. The College of Veterinary Medicine collects animal bones, hairballs and bezoars, the Entomology Department has been gifted with rare stick insects and ornate termite sculptures . . . in fact, every department on campus that interacts with the public has its own tales to tell of unusual donations, many of which they've proudly kept and regularly use in their teaching.

Not all such offers are of tangible things. The university is showered with ideas, too. As North Carolina's great land grant institution, NC State not only attempts to educate the students who fill its classrooms or to serve as a repository of facts and knowledge, but it also acts as a great clearinghouse of thoughts, schemes and possible innova-tions generated by citizens across the state. While these often turn out to be practical or useful enough to dis-seminate, many are so out of the ordinary that they help define the very edges and corners of the "box" that their originators are thinking so far out of.

The Gregg gets these, too. One afternoon toward the end of 1996, only a few weeks after Hurricane Fran had left a wake of destruction across the eastern half of the state, I received a call from an elderly man who said he lived only a few blocks from the campus. He then told me he had been disturbed by all the downed trees and limbs that still littered so many yards and sidewalks in Raleigh, but thought that they might present an unusual opportunity. "With all the unemployed dwarves and midgets in this world—they can't *all* be in the movies, you know—you'd think some-one would have come up with a midget sawmill that could put all those little people to work sawing up limbs and twigs and making small stuff like pencils, instead of just carting it all away to the dump. I've a good mind to do it myself!" I was intrigued (and amused) enough by his idea that I let him go on. After describing his midget sawmill scheme in further detail, he began to elaborate on other ideas for inventions, interspersing the descriptions with complaints about how he had been cheated out of fame and fortune time and time again.

"I was the one who came up with the idea of putting glue in a toothpaste tube," he claimed. "And ornamental curtain rod rings—that was my idea, too! And what about the triangular what-not shelf, designed so it could fit in a corner without jutting out into the room—that was mine! And now you see them everywhere!"

He was just getting warmed up. "Now here's one of my best ideas," he said, "but you can have it as a freebie. What about putting unemployed folks to work quarrying down in the bottom of the Grand Canyon? That would solve *three* things at once. Just think about it! It would give them-all a job, which would help solve unemployment. Meanwhile the Grand Canyon would just keep getting grander and grander! *And*—they could sell softball-sized pieces of the rock for souvenirs! It's a doggone win-win-win! Wouldn't that be something? You can take that to the bank!

"Tombstone novelties, decorative cabinet edges, using oyster shells as ashtrays . . . heck, there's a whole world of things I've thought of! I get so mad every time I go into a Wal-Mart and see *my* ideas and *my* inventions a-profiting everybody else but me. But this time I'm going to whup 'em all—nobody's going to beat me to the midget sawmill! You just watch!"

I could see he was getting "whupped" into a crank frenzy, and realized it might be getting to be time to gently terminate the call. To calm him down, I asked for his address and told him I'd soon be getting back in touch. A week or so later I followed through by sending him a letter on museum letterhead, telling him how much I had appreciated hearing his ideas. This had the effect of touching off a one-way torrent of correspondence that lasted for months. Sometimes two or three envelopes arrived in a single day, filled with descriptions and sketches of all kinds of ideas. Before long, I was receiving drawings that not only further developed the midget sawmill concept (with apologies that, "I realize this is only a rough sketch, but your engineers can make the professional details"), but other plans for inventions like sponge mops that could clean windows in high-rise buildings "without a need for scaffolding," techniques for "decorating the plain end of any house" by using curly iron rods, and moneymaking proposals involving the manufacture of miniature bricks stamped with "Souvenir Brick of the Capital of N.C." which could be sold "for $3.95 plus $3.00 shipping and handling, or $2.95 wholesale to gift shops around the state."

As the months marched on, the flood of ideas he sent me continued, with proposals for countertops protected with coin mosaics; snap-on plastic caps to put over the ends of braided hair "to keep it from becoming unraveled;" heart-shaped purses (so that loose coins would migrate to the point at the bottom where they could be easily retrieved); "bible mirrors" made of thin, polished metal that would not only serve as bookmarkers but also encourage the dedicated reader "to reflect on the state of his soul while reading;" cigarette lighters equipped with sliding, wind-proof shields to eliminate the need for cupping the hands; bike-spoke decorations shaped like squirrels or guitars; hunting cabin kits; snake-proof spats; special clippers for turning old newspapers into placemats ("Never need to wipe the table again!"), and copper necklaces made to resemble oversized Lincoln pennies (to be priced at $3.95, like nearly everything else he described). The sheer range of his creativity was impressive, even if his approach was scattershot.

After two years of being entrusted with the multitude of "breakthroughs" he revealed in these letters, I received one in which he announced, "The inventions have slowed down some now. I just got a letter from the Arts Council acknowledging my grant application. Toodle-oooh, Your Friend, the ole Inventor-Carpenter-Builder." I don't know whether he no longer needed my encouragement, whether his application succeeded or not, or whether someone at the Arts Council now became the new focus of his creative output (if so, I hope they were friendly) but in any case, I never heard from him again.

Other departments can cite equally unusual examples of how they had been challenged to deal patiently with the flow of ideas submitted by one or more inventive and enthusiastic citizens. Al Boyers, who has spent years in NC State's Industrial Extension Service (an outreach program of the College of Engineering), has saved many fine examples of correspondence from eager inventors. Among them, for example, is a proposal submitted by one man for a generator that could power hydraulic pumps that would supply turbines that in turn would spin the generator ("If you think it will work, please do not say anything about it to anyone," the breathless discoverer wrote, ". . . till we can get together and report it to the TV for publicity so that I can gain credibility with the bank to get a loan to get the whole thing patented"). Unfortunately, this classic example of perpetual motion would have violated one or more of the same Laws of Thermodynamics that have so far kept everyone else from succeeding. Friction, gravity, and mechanical inefficiency are likely always to keep perpetual motion from working, as Mr. Boyers explained in his measured response. His tone was polite and thorough, even though I am sure he felt like he was treading yet again down an old path worn deep by generations of cranks and nutcases.

A design sent to Boyers by another backyard inventor detailed a plan for "new free energy" to be provided by "burning sulphur with a mixture

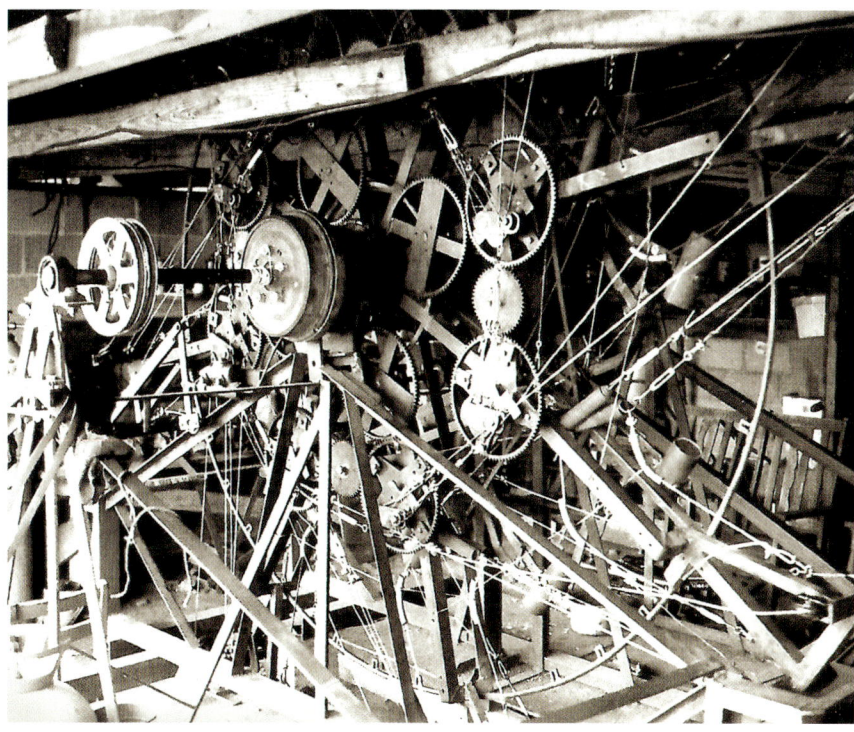

of corn meal in a [volcanic] crater" (the corn meal, according to the writer, would "help keep the lava from sticking to the flues"). Yet another file deals with an idea for a form of "Safe and Rapid Transit" that would utilize "revolutionary power plants for trains, ships and cars" and that could yield "48 billion horse power per minute from a 2½ h.p. lawn mower engine." It would, in fact, deliver enough power, so the man claimed, "to melt the legs off a pot belly coal stove in every [electrical] substation and even [melt] the tacks in the men's shoes." A snapshot was enclosed to prove that such a device actually existed, although there had been no reports of vast amounts of power emanating from the rural area where the letter originated.

Perpetual motion machine built in Orrum, NC farm shed, 1974. From the files of Al Boyers, courtesy of John Crow.

Despite such colorful imagery and outlandish claims, the hearts of such inventors are often in the right place. The originator of the last-mentioned concept continued his letter with the reminder that, "If this engine is used, then no more pollution to our streams from the oily coal waste of steam plants, no more nucular [sic] waste to poison the air, no more gas waste to kill the people and animals." Perhaps recognizing that once in a while a seemingly farfetched idea may hold the kernel of a real improvement, it is much to the university's credit that the majority of its faculty and profes-

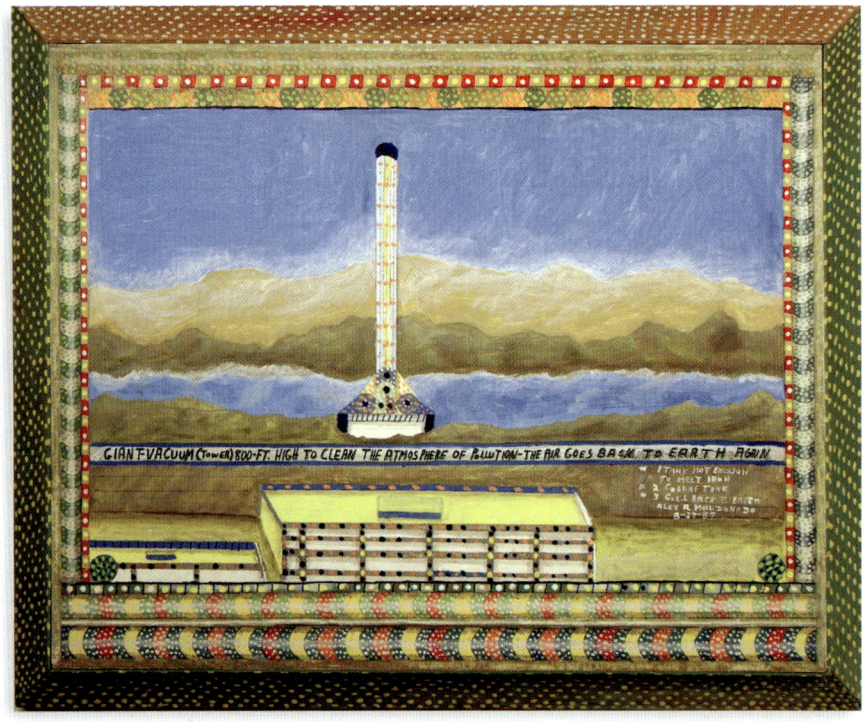

GIANT-VACUUM (TOWER) 800-FT. HIGH TO CLEAN THE ATMOSPHERE OF POLLUTION-THE AIR GOES BACK TO EARTH AGAIN

Alex Maldonado,
*Giant Vacuum
(Tower) 800-ft.
High*, 1987, 26
x 32 in., oil on
canvas board,
painted frame.
Courtesy of the
Ames Gallery,
Berkeley, CA.

sional staff try to respond thoughtfully to the taxpayers who make their careers possible, and do them the honor of taking them seriously.[2]

Also among Al Boyers' bulging files is a suggestion submitted by a retiree for a plan to generate power from moving vehicles by driving them up onto handy nearby ramps instead of applying brakes whenever they might need to come to a stop. According to the plan, the ramps would not only

2. Not all far-out concepts originate outside the university, either. Dr. Edwin H. Paget, a professor of rhetoric at NC State for almost 40 years, claimed to have received widespread recognition as "History's Most Significant Man" for his unusual ideas, not least of which was his belief that it would be possible to live to 140 or more merely by climbing stairs for several hours every day (he said he averaged 340 trips up and down his basement steps daily). His other innovations included a dieting device that hydraulically connected a chair to a table to encourage eating less (the more one weighed, the more quickly the table would rise out of reach); strobe lights attached to jeans to announce the wearer's needs (like "looking for lost kitten," "need assistance in home furnishing department") instead of advertising already-famous designers; colored "Paget Lights" that would allow homeowners to color-coordinate their dwellings with the changing seasons; goalie-free soccer (more scores); adjustable-height basketball hoops (fairer for short people); and annual brain tests for political leaders. On second thought, perhaps that last idea is not so far-out after all.

slow and stop the vehicles, but could slowly lower them back to ground level via a series of gears and counterweights that would spin flywheels attached to generators that would create electricity. A school, the inventor suggested, could be powered by the force of gravity acting on its own school buses, which would descend on individual ramps while disgorging their cargos of children. Somewhat smaller ramps placed at intersections below traffic signals could power the signal lights themselves. With such a ramp mounted in the family garage, the energy needed for cooking dinner might even be generated by the sheer weight of the breadwinner's car whenever he or she returned home from work.

With pages of detailed diagrams and equations, Boyers patiently responded that any energy "created" by the descending vehicles would be more than offset by the energy needed to propel them up the ramps in the first place, not to mention that the cost of constructing complex folding ramps everywhere one might conceivably need to stop a vehicle was probably prohibitive. Thus the proposal for "Gravitational Power Generation From Descending Ramps" was likely to be impractical. And yet, there is a kernel of a real idea here. Even if the would-be ramp inventor hadn't hit upon a more elegant way of channeling it, now, almost thirty years after he sent that letter, many of today's hybrid cars are recouping energy from slowing vehicles by using their momentum to generate more power instead of wasting the energy as useless heat.

Many of the artist-innovators celebrated in **FARFETCHED—Mad Science, Fringe Architecture and Visionary Engineering** intuited their way toward problem solving in much the same way. In one of his vivid paintings, retired prizefighter Alexander Maldonado depicted a "Giant Vacuum (Tower) 800-ft. High to Clean the Atmosphere of Pollution—the Air Goes Back to Earth Again." At the time he made this painting, such an idea was probably considered a bizarre and impossible (if not downright silly) concept. But today, decades later, serious scientists are exploring ways to permanently store CO_2 in underground geological strata, hoping this might impact climate change. Carbon Capture and Storage (CCS) is beginning to be a major field of research. Meanwhile, Paul Laffoley's idea for growing buildings from genetically altered tree seeds doesn't seem nearly as kooky as it did when he first proposed it almost twenty years ago. Such a technique—which, thanks to genetic engineering, now seems within reach—might make it possible to plant and grow shelters on other planets instead of hauling bulky raw materials clear across space to build them.

Even questionable ideas may be worth reexamining. For instance, now that recent studies are beginning to reveal the entirely unexpected beneficial efficacy of placebos—which often seem to work as well as "real" pills *even when patients are told openly that they have absolutely no medicinal value*—it may be time to revisit radionics, a healing system that relies on the power of intent as much as on the scientistic gear developed to channel it. Promoted in the early decades of the 20th century until it was deemed a "pseudoscience," there may be something to the idea behind radionics after all. Even phrenology (an idea long ago dismissed as quackery after it fell into the hands of eugenicists) may also be experiencing a resurgence, since the latest neuroscience is revealing that various emotions and abilities really *do* seem to lodge in localized "addresses" in the human brain.

It would be possible to keep on listing *ad infinitum* great ideas that were once thought to be crazy and crazy ideas that may someday turn out to be great, but perhaps the point has been made. What this all suggests is that human technology may be subject to much the same principle as biological evolution. Just as the vast majority of random mutations that occur as organisms reproduce may seem individually to have little measurable impact on how species differentiate and develop—but collectively make all the difference, and have led to all of creation—the vast majority of ideas, inspirations, experiments, and avenues of research may seem fruitless or unimportant, too. For every major breakthrough, there are millions of ideas that never go anywhere. If this starts to seem depressing, it's worth recalling Thomas Edison's famous remark that, "If I find 10,000 ways something won't work, I haven't failed . . . because every wrong attempt discarded is another step forward." In other words, the seething world of random ideas is entirely natural.

Meanwhile, until the works under consideration here are revealed as actual advances, it is already enough that they can be appreciated as art, and as wonderful examples of the creative imagination. As Albert Einstein once wrote, "Imagination is more important than knowledge. Knowledge is finite, imagination encircles the world."[3]

3. In 1932 Einstein also wrote, "There is not the slightest indication that nuclear energy will ever be obtainable. It would mean the atom would have to be shattered at will." Even recognized geniuses are often wrong, see.

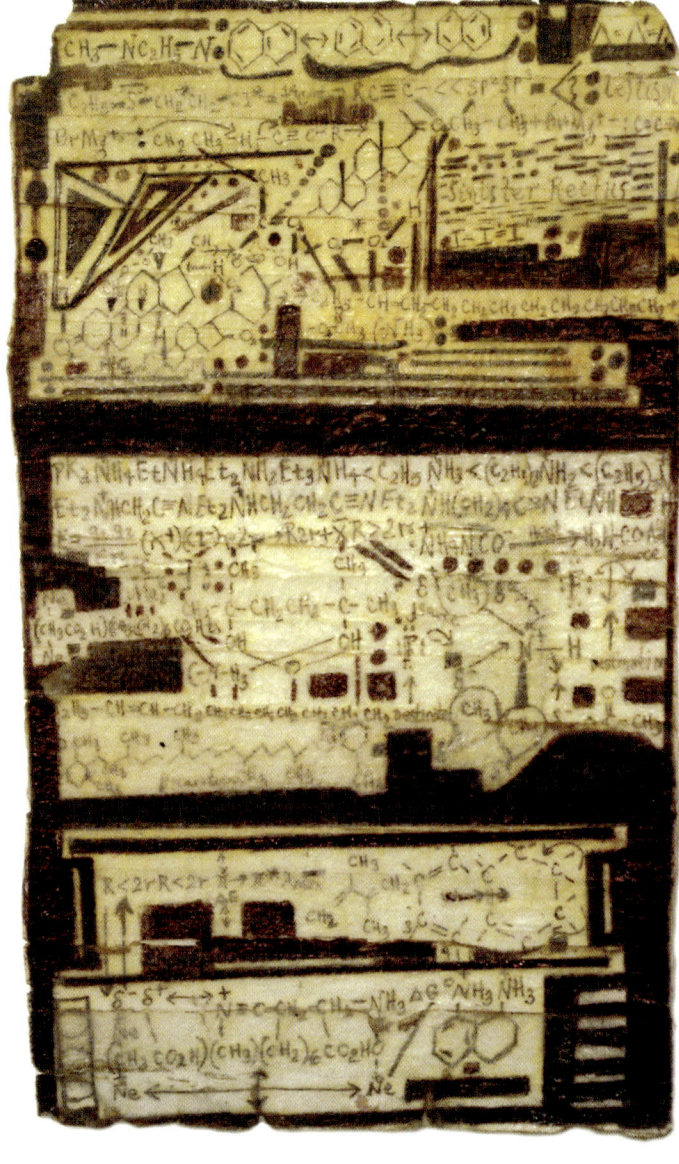

Melvin Way

Melvin Way was born in South Carolina in 1954 and moved with his family to New York City in the early 1960s. Like his four brothers, Way was musically talented; he played woodwinds and percussion in his high school orchestra, and later performed in various jazz and R&B groups. After leaving school, he worked as an industrial machinist and chauffeur, but recurring mental problems gradually left him with a sense of isolation and dislocation. Since the 1970s, Way has lived in homeless shelters and SROs (single-room-occupancy housing). An instructor in a hospital art class saw something in his cryptic schematic drawings and equations and brought them to the attention of an art gallery, which has begun to represent Way's work.
[Source: the Gallery at HAI]

Sinister Rectus, n.d., 5 x 3 in., black ink.
Courtesy of the Gallery at HAI, New York.

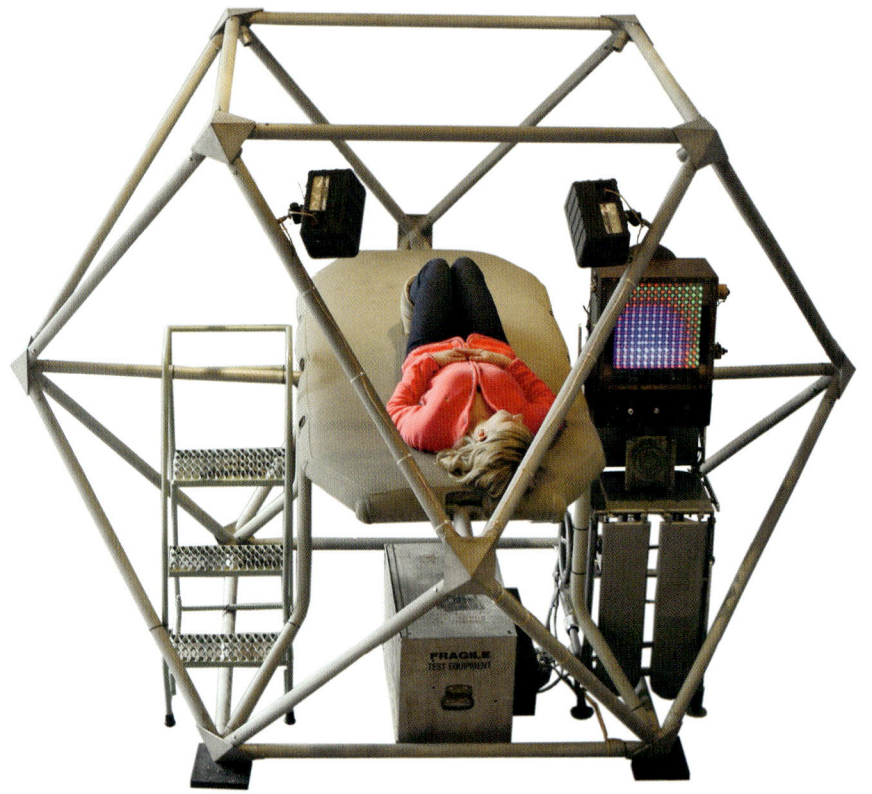

Duncan Laurie

Duncan Laurie was born in Detroit, Michigan in 1947, and spent summers on an island peninsula in Rhode Island. As an adult he manufactured belt buckles in Pennsylvania, taught art to juvenile delinquents, and worked as a traveling sandblaster.

Laurie's first attempts at art-making were fabricating his visions into glass paintings, writing journals on love's labor lost, and composing essays on the subject of implausible technology. Once he learned about electronic devices that could respond to directed human intent, he set about creating such a device as art. Through the good graces of the late R. J. Reynolds III, he obtained America's first immersive sonic radionic device. Based on the Buckminster Fuller dymaxion form, this *Betar* (an acronym for "Bio-energetic transduction-aided resonance"), was more familiarly referred to as the *Cotyledon* (after the first leaf to emerge from the seed of a flowering plant). Laurie made improvements and called it the *Purr Generator*.

By 1988 the *Purr Generator* had morphed into a unique kind of vehicle— the intended transportation method being entirely cerebral. Throughout its existence, various gifted basement inventors have repeatedly torn it apart and rebuilt it to new specifications. The latest version is powered by a sonic device mimicking the purr of a cat, designed by electrical engineer Gordon Salisbury, with V-J Todd Thille (a.k.a. Synesthete) providing a digital LED interface. Laurie has helpfully added a radionic Wishing Well for *Purr Generator* users who believe that such a machine can also work magic. [Source: Artist website, and author interviews with the artist]

....

Purr Generator (with Gordon Salisbury and Todd Thille), 2002 and 2011,
85 x 118 x 118 in. Mixed media, aluminum pipes, electronic circuitry, LED display.
Courtesy of Duncan Laurie.

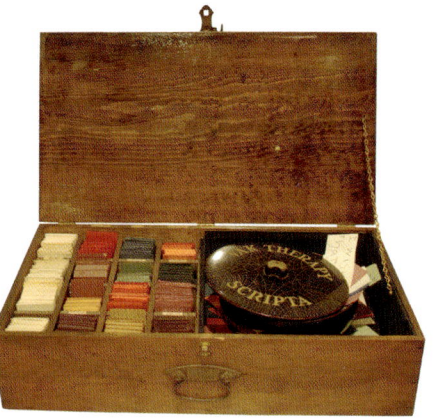

VAK THERAPY

It is not possible to say with any accuracy by whom, when or where this whimsical therapeutic device had its origin. To use it, one selects one of the brightly colored pieces of construction paper in the box and writes a wish or intent upon it. The wish is then placed in the specially made bowl or "Scripta." That was all that was required. The results were then left to one's Higher Power to be granted or denied. The universal enjoyment of Wishing Wells and similar radionics devices of all types and sizes is ample testimony to their recognized effectiveness, therapeutic or otherwise. [Anonymous maker, n.d., 2.75 x 16.25 x 9 in., wood, plastic, paper]

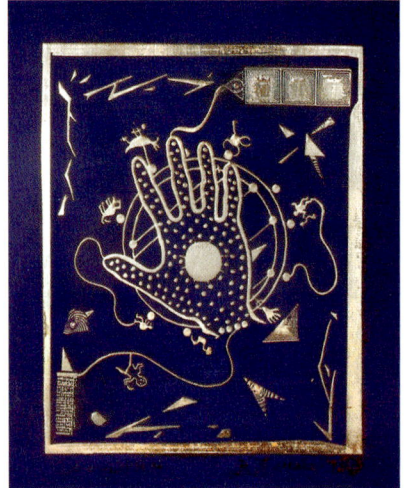

RADIONIC HAND

Outlines of handprints are often found among prehistoric cave paintings, suggesting that the hand provides a kind of signifier for the maker. In this cyberglyph, the silver outline of a hand containing a dot, connected to various symbols and graphics, shows the potentially magical linkage of the observer to the observed. If the image of the human hand asserts metaphorical control through this linkage, then this could be considered a radionic transaction as well. [Duncan Laurie, 1993, 7 x 6 in., printed circuit]

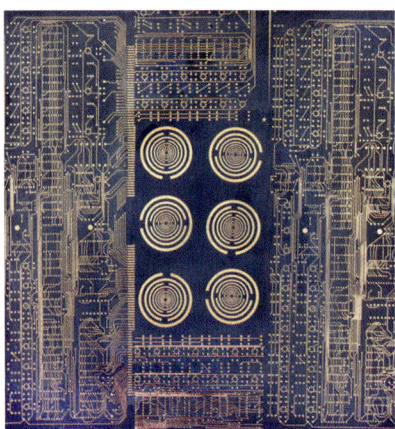

MULTI-WAVE OSCILLATORS

This cyberglyph depicts Multi-Wave Oscillators (MWO) grouped within random circuitry. The MWO are radionic resonators designed originally by Russian-born engineer and inventor George Lakhovsky between 1932 and 1942. They serve as a bio-electric healing apparatus that uses high frequency waves. In addition to his early radionic healing devices, Lakhovsky is now best remembered for his book, *The Secret of Life*, first published in France in 1929. [Duncan Laurie, 1994, 17 x 16 in., printed circuit]

CYBERGLYPHS

Cyberglyphs are akin to petroglyphs, symbols created by ancient peoples to act as signifiers for energetic transactions. This transaction nominally required natural energy in some form to expedite the human task. Just as the symbols on the surface of a printed electronic circuit board direct and influence the electrical currents passing through them, petroglyphs or earthworks have been thought to modify and direct telluric currents within the landscape towards a predetermined end. The cyberglyph images shown here are made of printed circuit board, some with silver plating and ink coloration.

RADIONICS

Radionics usually refers to a class of diagnostic and treatment instruments (and their uses) derived from the study of electronics in the early 20th century. Radionic instruments were originally designed to enhance and complement physicians' intuitive skills and to perform basic procedures like percussing or tapping the body with the fingers to ascertain the health of underlying organs. According to early radionics practitioners, healthy persons had certain energy frequencies that defined health, while unhealthy people exhibited other frequencies that indicated disorders. Radionic devices applied appropriate frequencies to correct the discordant frequencies associated with illness.

Radionics used the term "frequency," not in its standard meaning but to describe an alleged energy type that does not directly correspond to properties of energy described in mainstream physics. It's difficult to understand, and the devices are often legally prohibited from being employed on humans for medicinal purposes, but it's gaining in popularity nevertheless. Today, radionics technology is in widespread use throughout the world, especially among veterinarians.

Over time, as radionic methodologies have become more and more self-referential, they're no longer considered strictly scientific and practitioners have developed forms never imagined by its founders. Electronics are now combined with seemingly mysterious activities like dowsing to diagnose and treat diseases through the application of "directed intent."

Historically many cultures have invoked this same radionic principle through creative ritual: drumming, dancing, employing talismans, sand painting, and making rock art all serve to direct intent for specific purposes. Some applications are curative, others are agricultural; many are spiritual. Another offshoot is radionics as art. In this broader context radionics is only another means by which human consciousness seeks to know, heal and improve itself.

33

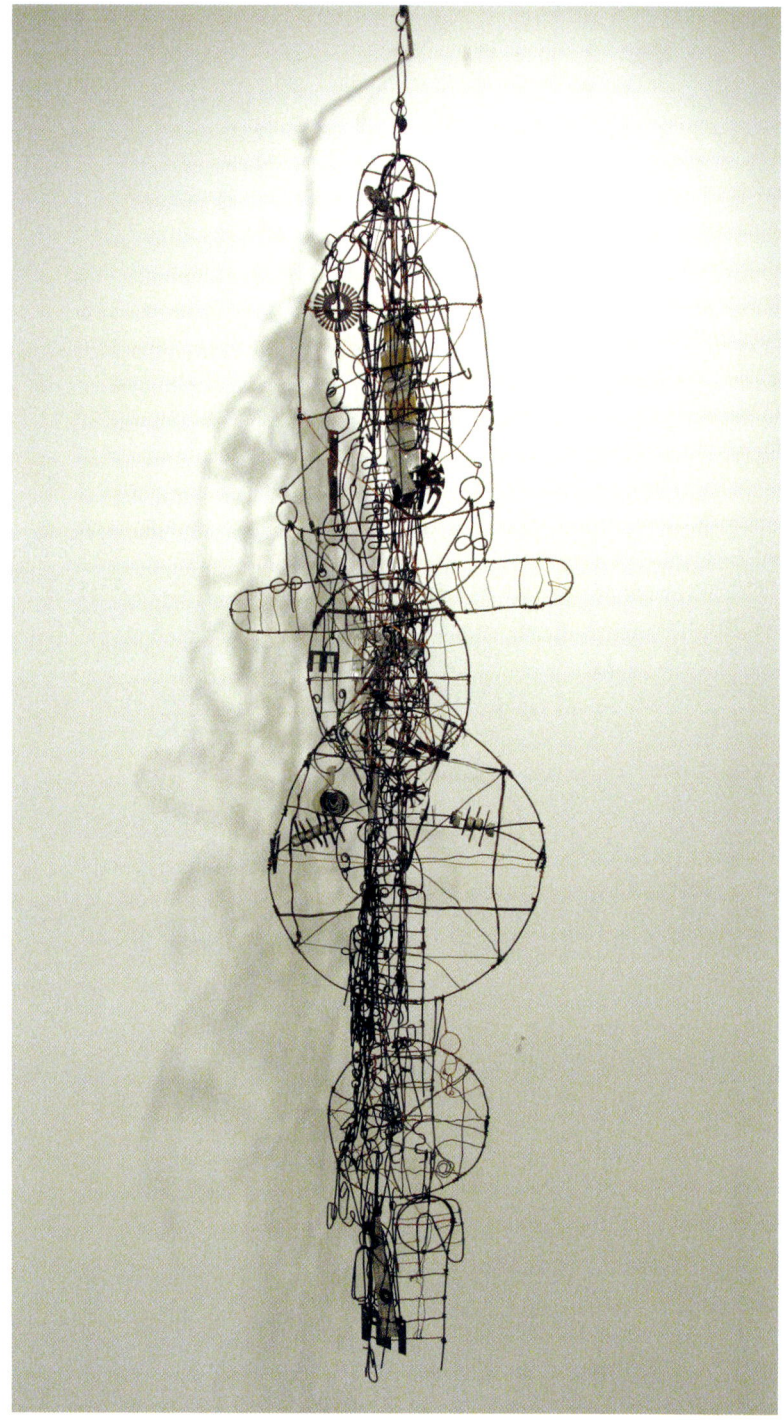

Emery Blagdon

Born in 1907, Emery Oliver Blagdon grew up in the Nebraska Sandhills, a region scoured by lightning storms and charged with static electricity from windborne dust. During a difficult childhood, both parents and a favorite sister died of cancer. It's unknown if Blagdon connected their deaths to electrical influence, but after he inherited an uncle's 160-acre farm near North Platte at the age of 48, he began converting an unheated shed and workshop into a mysterious indoor environment he eventually called his "Healing Machine." Blagdon intended to focus powerful electrical earth energies for curative benefits, since neighbors who suffered from pains or illness were often encouraged to enter the shed to feel the Machine's restorative electromagnetic power.

In the center of the shed was something like a model spaceship, with an elaborate wooden grid crammed with found wires, bottles, springs, and beer can tabs. This was surrounded by more than 400 other complex constructions resembling mobiles, chandeliers, television antennas and electrical generators, all of them made of salvaged copper wire, magnets, metal foil, containers of earth, bones, painted boards, waxed paper, and other substances and materials that could collectively help charge and heighten the device's energy.

Although many of the components looked seemingly haphazard, Blagdon frequently and exactingly reproduced them as multiples, suggesting that they were not abstractions or random symbols at all, but were intended as functional devices arranged like a circuit board.

A lifelong bachelor who never cut his hair, Blagdon worked on his machine every day for 30 years until his death in 1986. The bulk of what remains of Blagdon's strange device is now on permanent display at the Kohler Art Center in Sheboygan, Wisconsin. [Sources: Kohler Foundation, Inc., and John Michael Kohler Arts Center]

....

Component #337 of the *Healing Machine*, ca. 1960–80,
63 x 13 x 5 in., iron and copper wire, plastic, found objects.
Courtesy of Cavin-Morris Gallery, New York.

PERPETUAL MOTION

For most of human history, work has been accomplished only by sheer muscle power. Beginning around 8000 BC, domesticated animals came into usage for labor, followed some 4000 years later by the invention of the wheel (though not in the Western Hemisphere). About 300 BC the ancient Greeks built watermills, as well as windmills in the 1st century AD. But for the next 16 centuries, daily life for the vast majority of ordinary people remained more or less the same, with only tiny, incremental improvements.

The introduction of practical steam engines in the late 17th century was a leap forward. Instead of being limited to windy hilltops or waterfalls (when there wasn't a drought or freezing weather), mechanical energy could be achieved almost anywhere, anytime. This set off a frenzy of technological, scientific, and manufacturing progress—later termed the Industrial Revolution—marking the most important moment in human history since the domestication of plants and animals almost 10,000 years earlier.

However, the change wasn't all positive. Steam power demanded fuel to heat boilers, which led to intensive mining, deforestation, terrible pollution, low-wage labor, and a host of other problems. But if some kind of clean power source could be found to replace steam, none of that would be necessary. The search for fuel-less power quickly became (and still remains) the Holy Grail of industrial technology.

Enter Perpetual Motion. In 1670, only a few years after steam was introduced, Bishop John Wilkins, an official of the Royal Society (the Western world's first scientific organization), recommended that chemical extractions, magnets, or gravity might offer some way to generate constant mechanical power without the input of wind, water, or fossil fuels. Although these avenues later led to major discoveries like fuel cells, batteries, electric motors, and gravity-assisted space maneuvers, perpetual motion itself remained elusive. Researchers gradually came to believe that true perpetual motion—a machine that required no outside input of any kind—was probably impossible.

Nowadays scientists can point to the First or Second Laws of Thermodynamics as reasons why that is almost certainly the case, for every perpetual motion machine violates one or both laws. But even by 1775 the French Royal Academy of Science announced that it would "no longer accept or deal with proposals concerning perpetual motion" because they believed it was a waste of time. A handful of devices seemed nearly like perpetual motion. A clock in the Department of Physics at the University of Otago, New Zealand, for example, winds itself by changes in temperature or air pressure and hasn't been manually wound since 1864. But strictly speaking, since it requires external input from nature, it is no different from an ancient watermill and can't produce power on an industrial scale.

For the past 150 years few serious scientists have taken perpetual motion seriously, not just because it seems impossible, but also because the ability to transfer power over long distances via high-voltage lines makes it unnecessary. Hydroelectric power, tidal and wave-action generating stations, and geothermal, solar, wind and other energy sources that can be located wherever nature provides them, may not fit the classic definition of perpetual motion but may prove "perpetual enough" for most practical purposes.

Meanwhile the quest to achieve old-fashioned self-perpetuating motion has been left to cranks and eccentrics. And perhaps some day one of them will succeed? One might remember that many amazing achievements have been accomplished by people who did not realize (or refused to believe) that what they were attempting had been deemed impossible.

Harry Leroy Brunson

Harry Leroy Brunson spent 60 years attempting to perfect a perpetual motion machine, which he felt the U.S. Government could take advantage of. He worked in secret, using only a small handsaw and pocketknife in the basement of his home in Greenville, Alabama. The machine, which no one saw until after his death in 1991, depended in part on power derived from "devine [*sic*] intervention."

Born in Victoria, Alabama in 1898, Brunson was an eccentric but well-educated man who studied at Alabama Polytech (Auburn), UNC-Chapel Hill, and the University of Southern California. For work he owned a restaurant, ran a small feed store, and even claimed to have been a vaudeville performer, wearing pancake stage makeup even when not on stage. [Source: Author interviews with family]

....

Perpetual Motion Machine, 1940–1990, 60 x 102 x 84 in.,
plywood, lumber, steel shafts, lead weights.
Courtesy of the Taubman Museum of Art, Roanoke, VA.

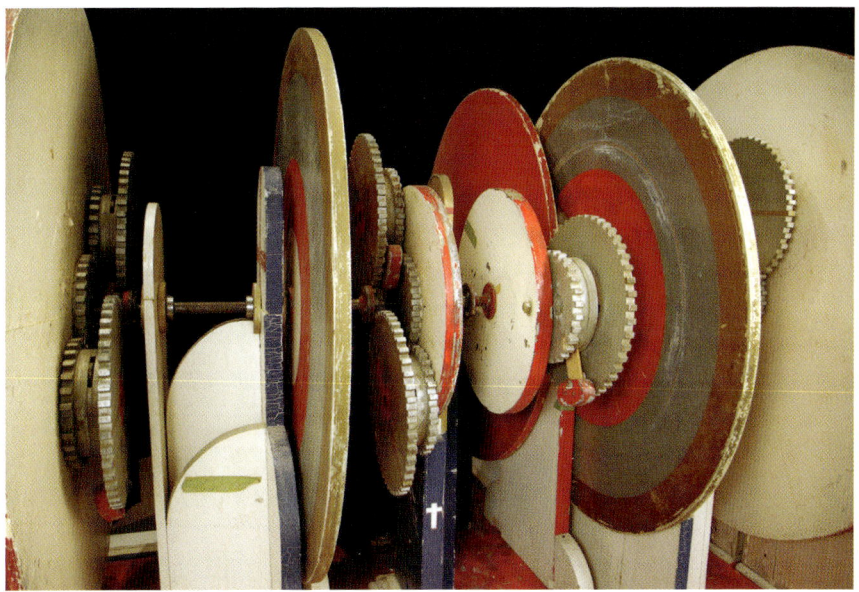

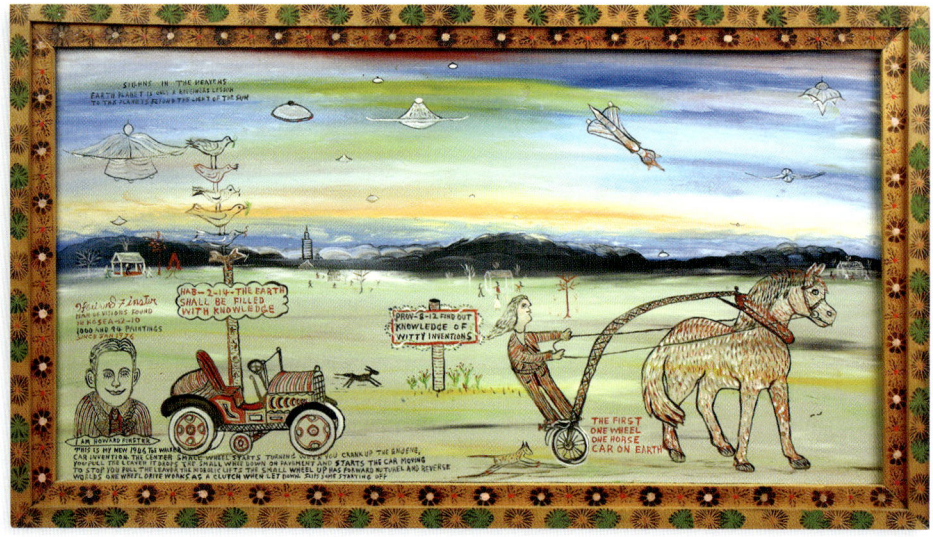

Find Out Knowledge of Witty Inventions, n.d.
(Finster number 1000 and 94), 17.5 x 31
in., enamel on panel, pyrographic frame.
Courtesy of Jane and Bert Hunecke.

The One Wheel Course Power Drive, ca. 1960,
12 x 9.25 in., ballpoint on graph paper.
Illustrates an idea that occurred to Finster in
1947. Courtesy of Jane and Bert Hunecke.

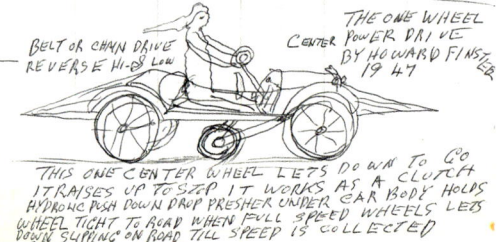

Howard Finster

Howard Finster was born on a small farm in northeast Alabama in 1915. Following a sacred calling while still in his teens, he preached the Gospel for five decades, meanwhile raising a family and working at a variety of other trades. His early talent for drawing and handicrafts led to part-time work as a sign painter and a purveyor of his own decorative scrap-wood clock cabinets with burned-on decorative patterns.

When he was in his 30s he built a cluster of miniature gable-roofed, multistory towers up to six feet tall in his yard. This fanciful environment became a prototype for his famed "Paradise Garden," a much more ambitious yard-art project undertaken in the early 1960s, after he moved to Pennville, Georgia. Finster drained a swamp behind his house and built a series of walkways, small buildings, bridges and sculptures of concrete embedded with glass, coins and found objects.

"Paradise Garden" featured various flowering and fruit-bearing plants, as well as his collection of the "Inventions of Mankind" which was intended to eventually include "one of everything in the world." Finster claimed he built the garden after a vision of a giant man briefly appeared at his gate, instructed him to "Get on the altar," and then disappeared. The garden is now owned by the Chattooga County government, which operates it and keeps it open to the public through a foundation that is also supervising restoration of the site.

In the late 1970s Finster began an increasingly prolific output of text-heavy, cartoonish paintings depicting spacecraft, impossible machinery, unusual architecture, biblical scenes, and American heroes (including himself among them). He said many of these paintings were inspired by "visions of other worlds beyond the light of the sun," a source he also credited for the elaborate makeover he gave to a one-story wood-frame building on the edge of his garden. He transformed this humble structure into the art-festooned "World's Folk Art Church," featuring a mirror-lined spiral staircase at the center of a vertically tapering, four-tiered tower that resembles a steeple-capped wedding cake.

By 1985 Finster had attained celebrity status in the overlapping realms of contemporary art and pop culture. He continued to make small paintings derived from his earlier work until his death in 2001. [Sources: Author interviews with the artist, and John Turner, *Howard Finster: Man of Visions*, (New York: Knopf, 1989)]

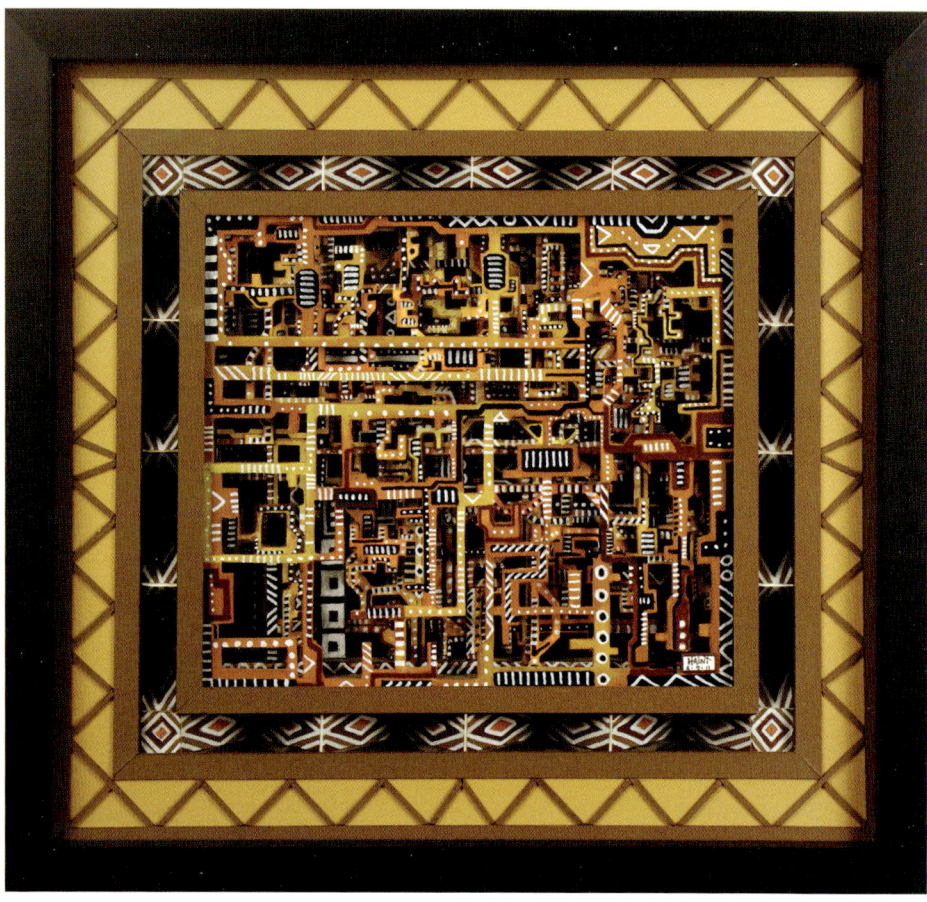

Alan Wayne "Haint" Bradley

Wayne Bradley was born in Red Bank, Tennessee in 1947 but grew up in Los Angeles, California, where he says he was influenced by the beatnik coffee houses and psychedelic art of the late 1950s and early 60s. He lived in Lafayette, Georgia, for 20 years where he worked as a truck driver.

About his work, he says, "I draw and make circles. The circles represent the fullness of God's creation. And the industrial style represents our high tech culture that is often at odds with His creation. God made the whole world in one great intricate pattern, and when we change things in it without knowing what we are doing we can ruin His pattern.

"I've made bad choices in this temporal life of mine. I believe that you reap what you sow. Love is the most important thing and the key element of wisdom." [Source: Jimmy Hedges, Rising Fawn Folk Art Gallery]

....

Tribal Circuits (opposite), 2011, 16.75 x 18 in. and *Fueling the Machine* (below), 2010, 16 x 31.25 in.; cardboard, metal, plastic, paint, wood.
Courtesy of Rising Fawn Folk Art Gallery, Lookout Mountain, GA.

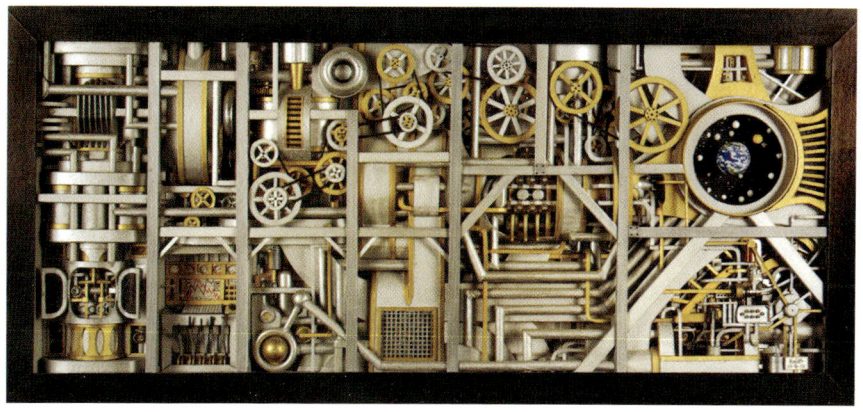

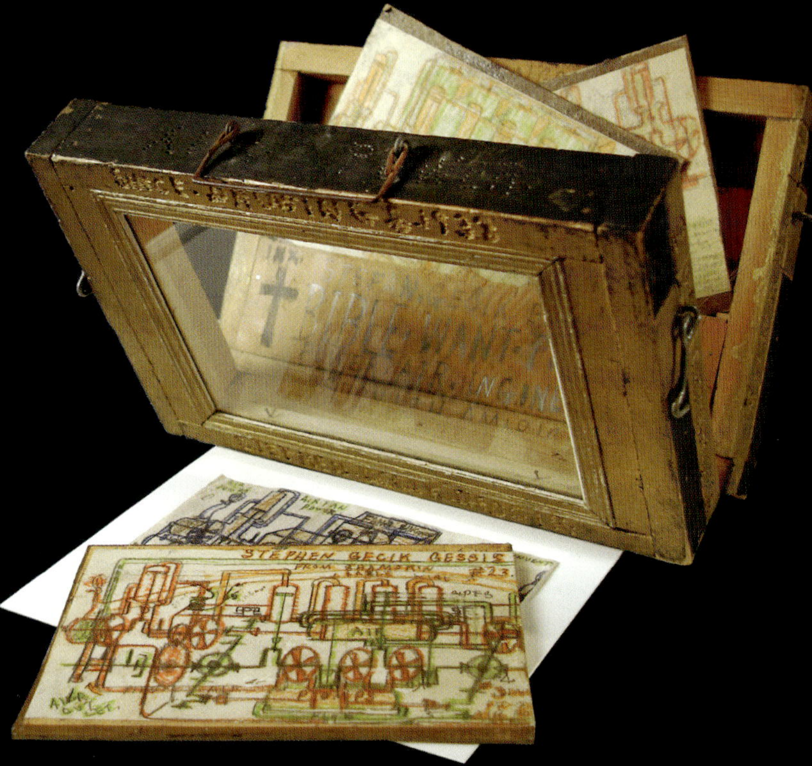

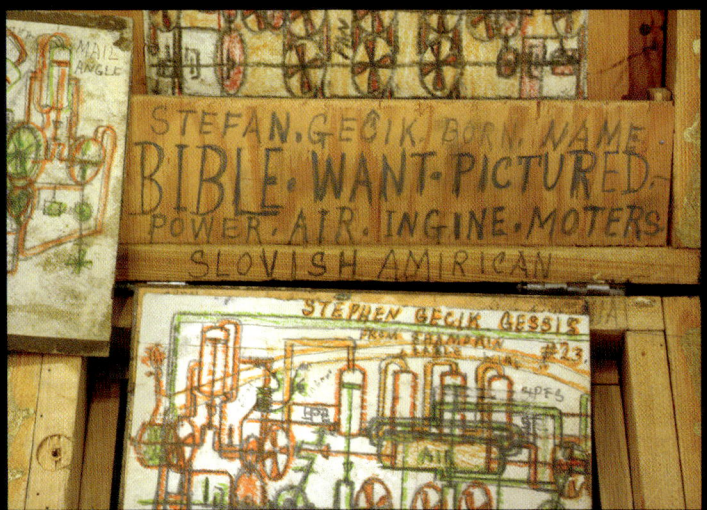

STEFAN·GECIK·BORN·NAME·
BIBLE·WANT·PICTURED·
POWER·AIR·INGINE·MOTERS·
SLOVISH·AMIRICAN

STEPHEN GECIK GESSIS
FROM TRUMPKIN
#23.

Stephen Gessig

Stephen Gessig (b. 1903-?) was an amateur inventor about whom little is known except for what can be gleaned from about 40 drawings and some cryptic notes in a wooden case discovered several years ago by an art dealer. Information he recorded in these documents includes several alternate spellings of his name and indicates that he lived in Shamokin, Pennsylvania. It also suggests he was incarcerated for a time in an unnamed state mental institution.

Gessig's drawings are diagrams of inventions accompanied by handwritten notes. All of this material was contained in a handmade wooden presentation case painted gold and incised with related texts. A window cut into one side of the case is fitted with a glass pane, and a corresponding window on the other side is covered with a wire screen, allowing two drawings to be partially viewed even when the case is closed.

Most of the drawings are double-sided and glued to cardboard. Some of them were originally found glued back-to-back, and have been carefully separated. Notes on the drawings include dates that range from 1936 to 1956. Inventions depicted and explained in the drawings include plows, kitchen gadgets, airships, a perpetual-motion car and—more ominously— a mechanically augmented, cross-shaped chamber for committing suicide.

[Source: Aarne Anton, American Primitive Gallery]

....

Inventor's Box, n.d., 9 x 15 x 3.5 in., wood, glass, screen, crayon and ink and board. Courtesy of American Primitive Gallery, New York.

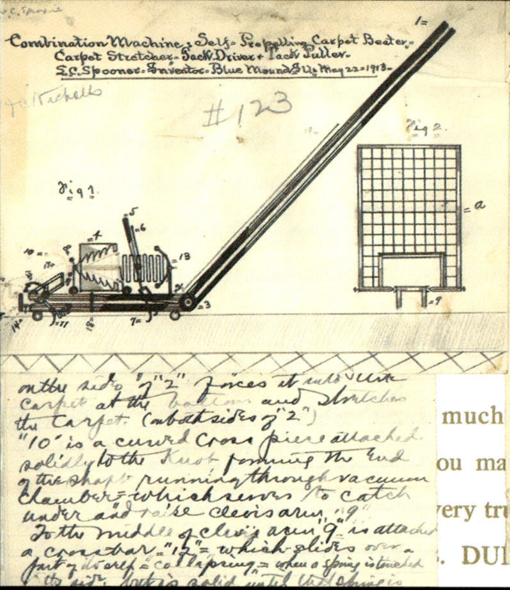

Combination Machine = Self = Propelling Carpet Beater =
Carpet Stretcher = Tack Driver + Tack Puller.
E.C. Spooner = Inventor. Blue Mound Ill. May 22, 1913.

#123

Fig 3.

on the side "2" forces it into the
carpet at the bottom and stretches
the carpet (what sides of "2")
"10" is a curved cross piece attached
solidly to the knob forming the end
of the shaft running through vacuum
chamber = which serves to catch
under and raise clears away. "9"
To the middle of clears arm "9" is attached
a crossbar "12" which glides over a
part of shelf = coil spring = when a stone is enrolled

"much
you ma
very tr
DU

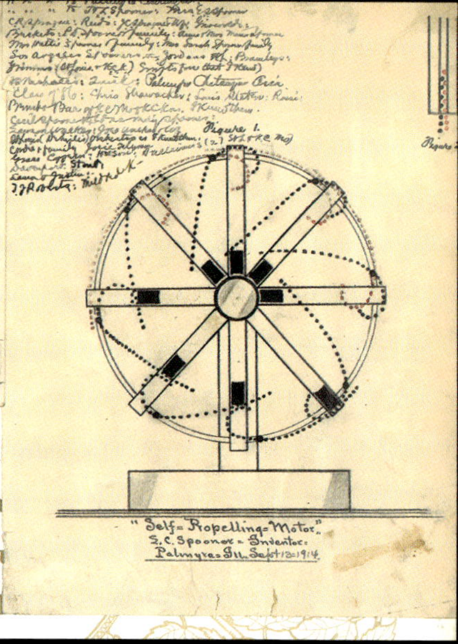

"Self = Propelling = Motor."
E.C. Spooner = Inventor.
Palmyra, Ill. Sept 13, 1914.

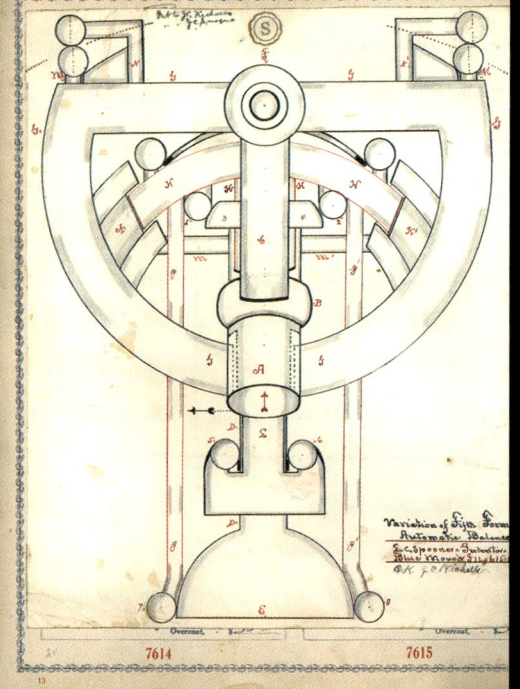

Variation of Life Form
Automatic Balance
E.C. Spooner = Inventor.
Blue Mound Illinois.
P.K. & O.K. Nicholls.

Overcoat   Overcoat
7614   7615

L.C. Spooner

Almost nothing is known about Lee C. Spooner other than what was discovered in a recycled fabric-sample scrapbook in which he had pasted hundreds of sketches for inventions. Dating from 1911–1935, the book is filled with proposed patents for "practical" designs using principles of perpetual motion and self-propulsion. But the practicality of Spooner's "self-propelled trash can," "self-propelling wheel-barrow," and "self-propelled finger-lifter" are difficult to grasp.

Unfortunately, the perpetual-power sources of his self-propulsion schemes seem to have disappeared with the man himself. References in the scrapbook suggest that Spooner lived in or around the Illinois towns of Decatur, Palmyra, and Blue Mound, but no other biographical details are known. In the upper part of some of the pages it is possible to read: "P.B. Spooner. DRUGS, JEWELERY, BOOKS STATIONERY, PAINTS, WALL, PAPER, SHADES, DRUGGIST SUNDRIES —PALMYRA, ILL." Spooner's family evidently owned a pharmacy.

One of the main things that sets Spooner's oeuvre apart from that of so many other inventors is that he seems to have taken for granted that perpetual motion would work. Others are preoccupied by their explanatory designs, but Spooner skipped the physics to focus upon what perpetual motion could be used for, seemingly presuming that it was on the verge of becoming available. As soon as it did he aimed to cash in with patents on everything it could possibly be used for, already in place.

Some of his ideas are obviously extreme. A miner's pick or baby carriage powered by perpetual motion would be annoying if not outright dangerous. But the Midwest, in Spooner's day, had such a rudimentary electrical grid that running a carpet beater or vacuum cleaner with an onboard perpetual motion machine probably seemed no more fantastic than accessing power for appliances via outlets and wires leading from dams located many miles away. [Source: Stephen Romano]

....

Self-Propelling Carpet Beater, 1913, 6.5 x 10 in.; *Self-Propelling Motor*, 1914,
10.5 x 7 in.; *Variation of Fifth Form of Automatic Balance*, 1914, 14.75 x 11.5 in.,
graphite and ink on found paper. Courtesy of Stephen Romano, New York,
and James Brett, The Museum of Everything, London.

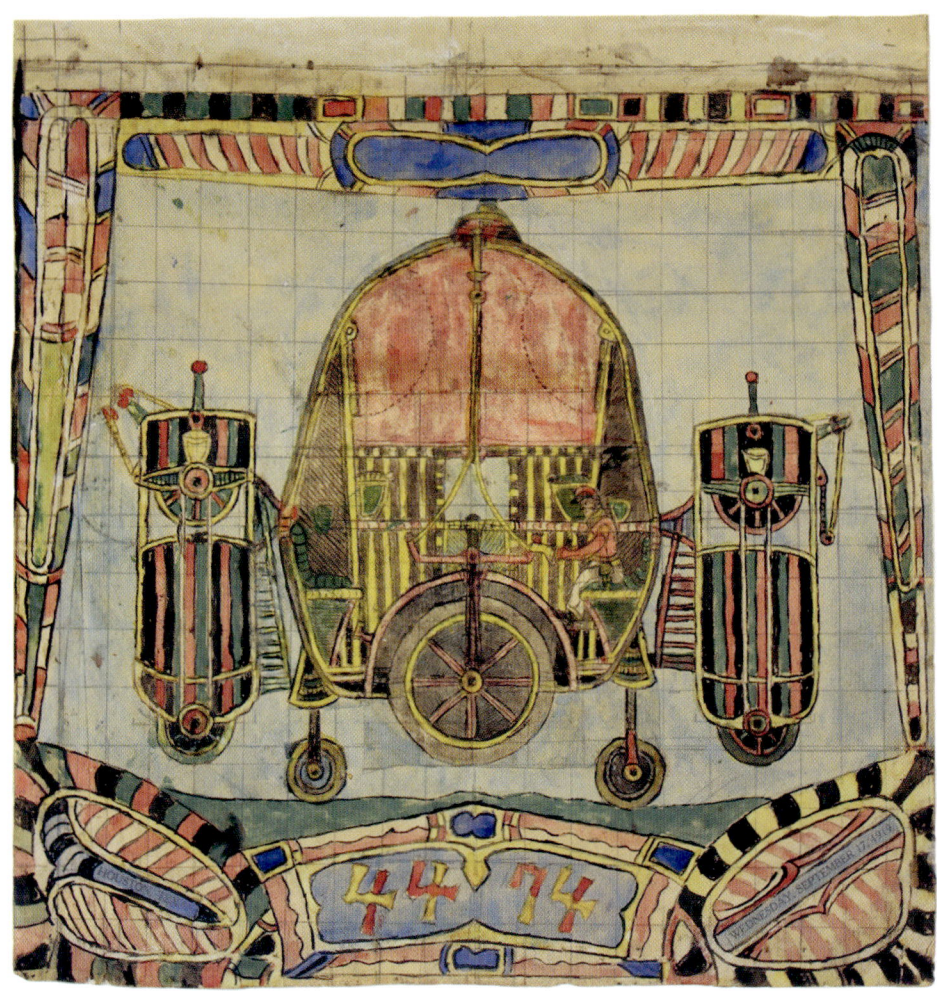

Charles Dellschau

Charles Dellschau was born in Prussia (later a part of Germany) in 1830, came to the U.S. not long after the American Civil War, and worked as a butcher and clerk at a German-speaking Texas community until his 1899 retirement in Houston. From then until just before his death in 1923, he drew and painted compulsively in his attic apartment, filling more than a dozen large notebooks with hundreds of drawings, watercolors and poster-paint images of airships and flying machines, collaged with clippings about the new aeronautical age ushered in by the Wright Brothers. Dellschau is now regarded as one of the nation's most important outsider artists.

His work was almost lost. Thirteen bound scrapbooks of his paintings were discovered in a Houston landfill and rescued by a furniture dealer. A student at St. Thomas University heard about them and asked to include them in an exhibit about early flight. From there, they came to the attention of major collectors and museums, and have been widely exhibited and collected.

While Dellschau's aeronautical notebooks featured drawings of inventions outfitted with flapping wings, pulleys, cranks and striped canopies, and look far more like balloons, dirigibles or flying carriages than the biplanes of early aviators, they reveal a deep obsession with flight in a myriad of forms.

Dellschau also wrote long descriptions of an organization called the Sonora Aero Club, purportedly a secret group of inventors and early flight hobbyists who had discovered an anti-gravity substance called "NB Gas" that could provide both lift and propulsive fuel for flying machines called "Aeros." Dellschau claimed to be the club's draftsman. In an encrypted story hidden throughout the drawings themselves is information about an even larger secret society of which the Sonora Aero Club was supposedly only a branch, called NYMZA. Unfortunately, other than these mysterious references in Dellschau's works, researchers have so far uncovered no traces of either the Sonora Aero Club or NYMZA. [Sources: Stephen Romano, and Cynthia Greenwood, "Secrets of the Sonora Aero Club: A tale of UFOs, art collectors and the shadows of history," *Houston Press*, 12/10/98]

....

Airship 44–74, 1919, 19.25 x 15.25 in., watercolor, poster paint, pencil, collaged news clipping. Courtesy of Stephen Romano, New York.

48

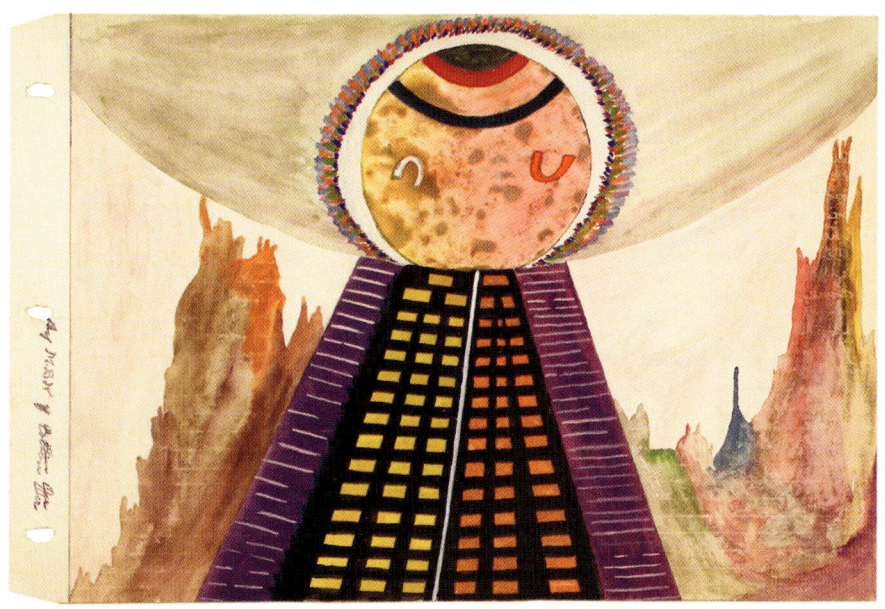

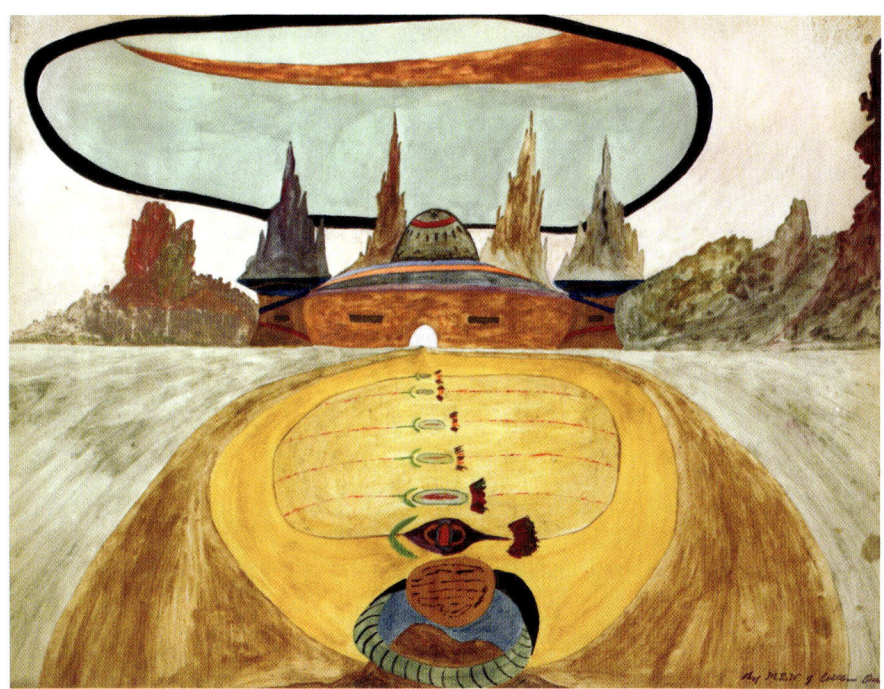

Melvin Edward Nelson

Born in Michigan in 1908, Melvin Edward Nelson was a family man. But in 1942, he abandoned his wife and children to move to Oregon, working as a general electrician telephone installer. He later bought 88 acres in the foothills of the Cascade Mountains near Colton, and moved there with his friend Cleo "Mac" McClintock.

Both ambitious inventors, they lived in separate cabins and worked much of the time. A barn served as a crowded storehouse for spare electronic and mechanical gear, while other ramshackle buildings were filled with Nelson's hoarded items. At some point Nelson stopped bathing, shaving, and cutting his hair. He believed that his property was on top of a vast, underground UFO base, and he claimed to have seen many spacecraft land there

In 1961 he began having cosmic and microscopic visions. Not only could he project himself into outer space to observe the Earth and the formation of new worlds, but he could also perceive these phenomena at an atomic level.

Nelson saw his ability as a divine gift for seeing scientific and spiritual truths, recording them as paintings on scavenged paper. The topographical-looking abstractions that he called "Photo Genetics" were made with pigments mixed by hand from rocks and minerals gathered on his property at night, including "stardust" from the landing and launch sites of UFOs.

Another series called "Sentra Photo Thesis" records what he saw through astral projection, aided by an invention he called an "anyzager." His other inventions included a "cyclotronic generator" and a "planetron," designed to generate cosmic energy, and equipped with a window through which, Nelson claimed, he could view outer space and receive extraterrestrial communications.

After a 1981 court battle over Nelson's property made him a ward of the court, he was institutionalized and administered antipsychotic drugs. He stopped working on his art and his inventions, and died in 1992. [Source: Peter Hastings Falk, "The Cosmic Visions of Ed Nelson," Madison, Conn: Falk Art Reference, 2007]

....

Futurist Visions, (opposite above) 11.25 x 16.5 in., (below) 17.5 x 24.25 in., graphite, poster tempera and watercolor on paper. Courtesy of Cavin-Morris Gallery, New York.

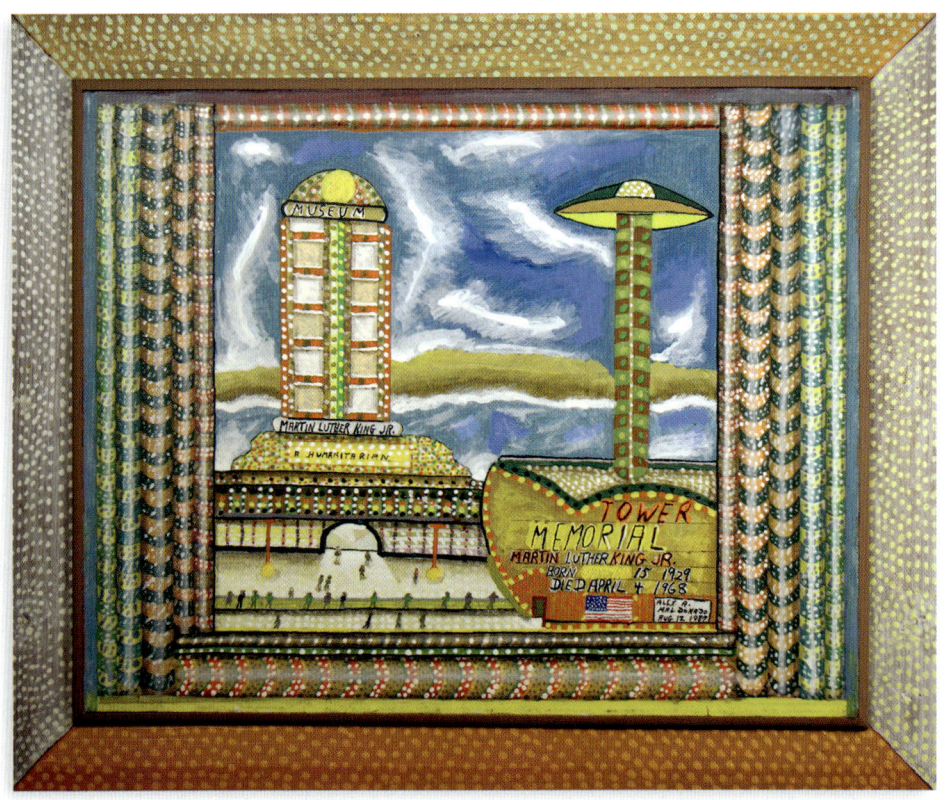

Alex Maldonado

Alexander Maldonado was born in Mazatlan, Sinaloa, Mexico, in 1901 and emigrated with his family to San Francisco in 1911. After his father's death when Alex was a teenager, he helped support the family by working as a shipyard riveter, also moonlighting as a professional boxer (a "left-hook artist," he once said) under the ring name of "Frankie Baker." Despite success in the ring, he gave it up to work in a cannery.

Maldonado never married and lived with his sister until her death, just four years before his own in 1989. At his sister's suggestion, he took up painting, choosing subjects from memory or his inventive imagination.

Maldonaldo's proposal for a *Giant Vacuum (Tower) 800-Ft. High to Clean the Atmosphere of Pollution—The Air Goes Back to Earth Again* presages recent proposals for ameliorating global climate change by storing CO_2 underground. [Source: the Ames Gallery]

....

(opposite) *Memorial To Martin Luther King, Jr.*, 1987, 27.75 x 33.75 in.;
(below) *21st Century—5 Telescopes In One—X-ray Computer*, 1986, 21 x 27 in.,
oil on canvas board, painted frames. Courtesy of the Ames Gallery, Berkeley, CA.

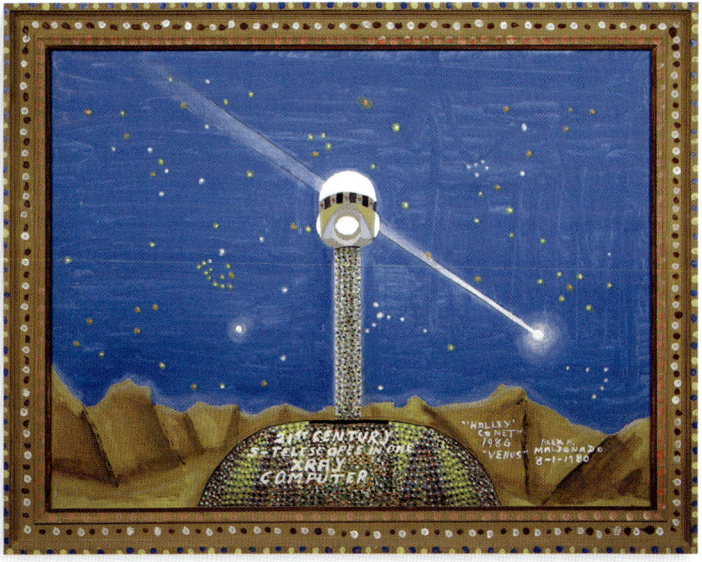

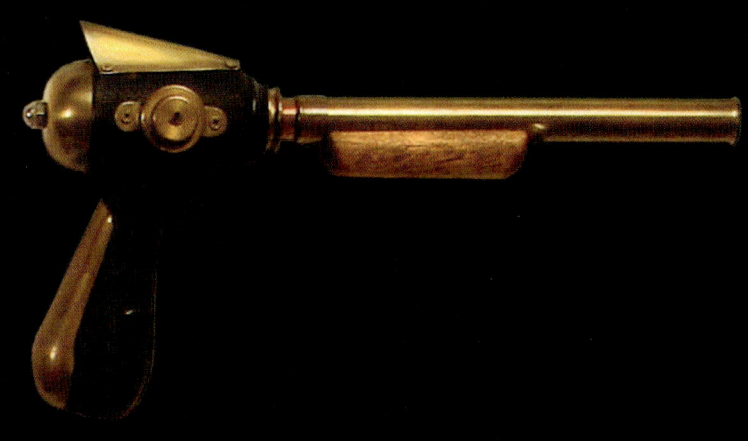

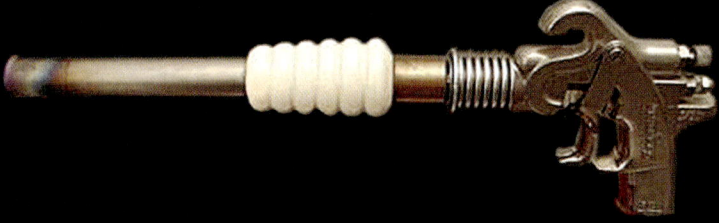

Clayton Bailey

Born in 1939, Clayton Bailey grew up in small Wisconsin towns. Known as an outgoing academic high-achiever, practical joker and mischievous backyard chemist, his formative influences included magazines like *Mad*, *Popular Science* and *Popular Mechanics*, as well as EC horror comics.

While studying chemistry at the University of Wisconsin, Bailey took an undergraduate ceramics class with Harvey Littleton, who introduced Bailey to the world of contemporary art and craft. Bailey began making clay sculptures and went on to get a master's degree in art and art education.

After teaching throughout the Midwest, he and his wife Betty moved to Porta Costa, California in 1968, where Bailey taught ceramics at nearby California State College, Hayward. During the 1970s he emerged as a leading figure in California's Funk movement, known for creating full-scale, ceramic settings such as the *Mad Doctor's Laboratory* and the *Wonders of the World Museum*, both nominally operated by his alter-ego "Dr. George Gladstone." Headquartered in downtown Porta Costa during the late 1970s, the "museum" housed exhibits including a full-size Bigfoot skeleton, a ceramic cyclops skull and an array of "fossil-testing equipment."

In addition to his ceramic work, Bailey created scores of robot figures and "ray guns" from an array of dismantled power tools and appliances. His lesser-known inventions include a repeating Whoopee Cushion and a novelty cup that squirts liquid in the face of anyone who tries to drink from it (for which he holds a U.S. patent). [Source: Diana L. Daniels, "Clayton Bailey's World of Wonders," Crocker Art Museum, Sacramento, CA]

....

Ray Guns, n.d., (opposite above) 9 x 13 x 3.5 in.; (below) 7 x 14.5 x 3 in., metal, wood, porcelain insulator, found objects. Courtesy of Dr. Stephen R. Turner.

WAR AND INVENTION

Many of the most significant advances in technology were first developed as weapons against enemies including, in early days, large predators. As copper implements first began to replace stone tools, the first items made were knives, spears, and battle axes, not hoes or shovels. Bronze helmets and swords were made before metal plowshares, too. When wheels were invented, war chariots preceded farm wagons as their first serious application.

The pattern still holds true today. Toys like the Slinky, Silly Putty, and Frisbees, along with practical implements such as tampons (originally developed as sanitary wound dressings), aerosol sprays, plastic, satellites, canned orange juice, rockets, jets, microwaves, computers, and the Internet are only a few examples of the thousands of spin-offs from the brainpower invested in war and defense. Military research is often backed by major financial support unfettered by concerns regarding economic viability.

Meanwhile, the finest minds, from Archimedes and Leonardo da Vinci to Werner von Braun and Albert Einstein, were often found in the employ of rulers or governments who focused their intellects on improving warfare. Besides painting the *Mona Lisa* and *The Last Supper*, Leonardo, for example, designed tanks, machine guns, mobile bridges, robot knights, trench pumps, wheel-lock muskets, water-cooled cannons, scuba gear, and defensive fortifications on behalf of his aristocratic warlord patrons. Working some 1,750 years before Leonardo, the Greek mathematician Archimedes is credited with developing catapults and other siege engines, ship-sinking machines, and a "heat ray" of parabolic mirrors that focused sunlight to set fire to the attacking Roman fleet during the two-year Siege of Syracuse (214–212 BC). The siege lasted so long largely due to the success of his defensive weapons.

"Death Rays" have since been one of the standard dreams of mad science, so much so that when California researcher Theodore Maiman proposed building the first laser (an acronym for Light Amplification by Stimulated Emission of Radiation) his employers discouraged him and wouldn't fund the project. Even after he went ahead on his own and succeeded, the American scientific journal *Physical Review Letters* refused to publish his paper on the subject, so he announced it in England instead. Major institutions such as Bell Labs, RCA, Lincoln Labs, IBM, Westinghouse, and Siemens had spent millions creating devices that weighed several tons and didn't work. Maiman's laser cost less than $50,000 in research and development, weighed only four pounds, and succeeded. He attributed his win—which has revolutionized communications, surgery, music recording, remote measuring, microscopy, supermarket checkouts and hundreds of other applications (in addition to many military uses)—to his "simplicity, unconventional thinking, and maverick spirit."

André Robillard

Born in 1931 as the son of a forest ranger who worked in the Forest of Or-
léans (60 miles south of Paris), André Robillard exhibited behavioral difficulties as a toddler. By age 7 he was sent away to a school attached to the
psychiatric hospital at Fleury-les-Aubray. There he did hard labor and farm
work like many other "feeble-minded" children during the Second World
War. After becoming increasingly violent and running away on multiple
occasions, Robillard was moved to the incarcerated section of the hospital
when he was 19.

Gradually his violent streak dissipated. By 1964, when Robillard was
33, he had settled down to the routine of hospital life enough that he was
placed in charge of the hospital laundry and sewage treatment plant. Living in his own apartment on the hospital grounds, he began creating hundreds of weapons and war machines as well as "sputniks" and spaceships out
of scraps of recycled wood, used light bulbs, tin cans and other materials
scrounged on the hospital grounds. Machine guns, death rays and exotic
rifles of all kinds fascinated him. When asked why he focuses on making
such seemingly violent devices, the now-gentle Robillard only smiles and
says, "to kill misery and suffering." [Sources: author interviews with the artist, and ABCD Collection]

....

Fusil x USA A.W.R. Rapide T136, ca. 1985, 22 x 45 x 8 in.,
wood, found objects, mixed media. Anonymous lender.

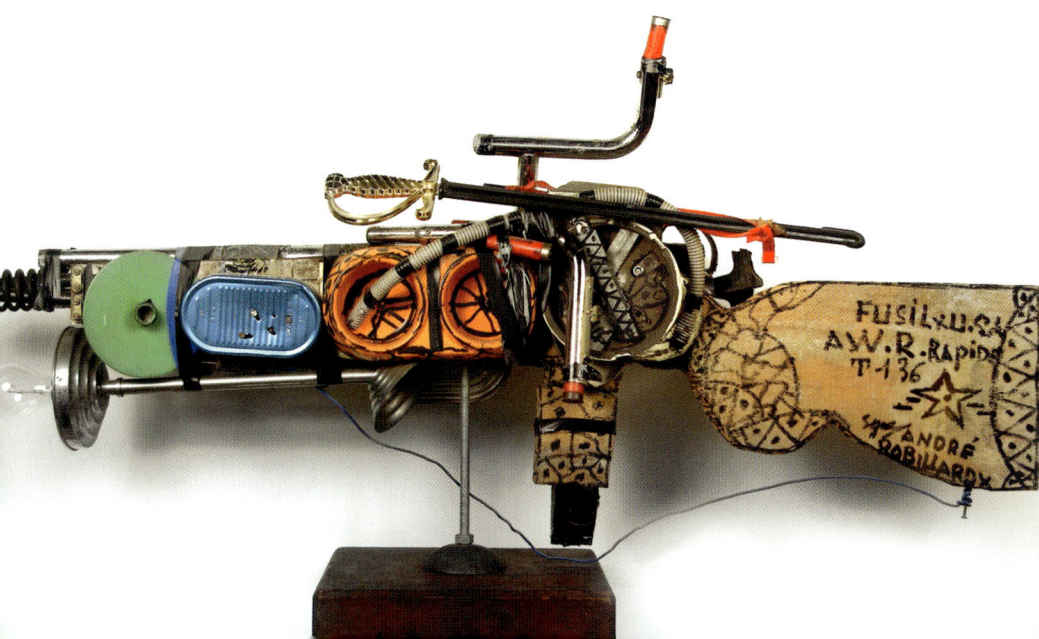

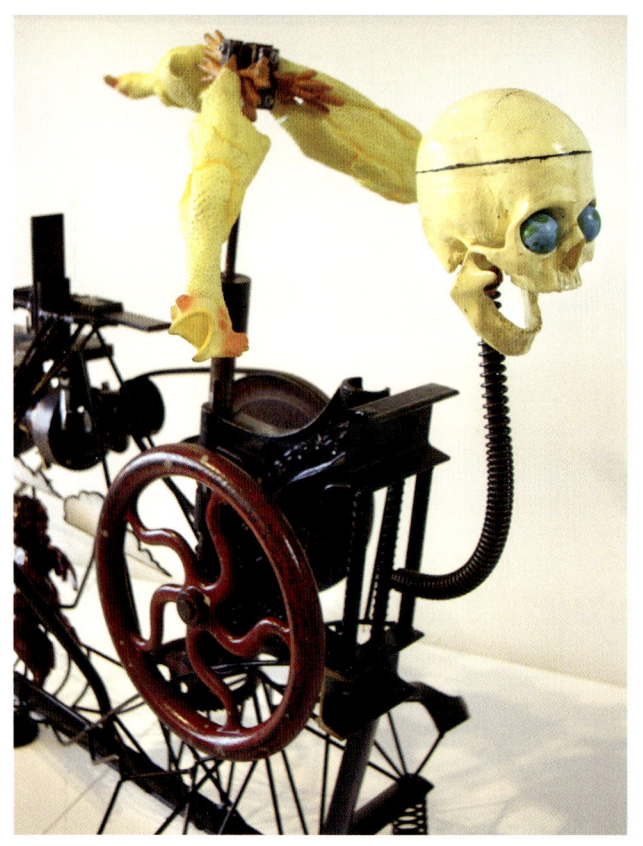

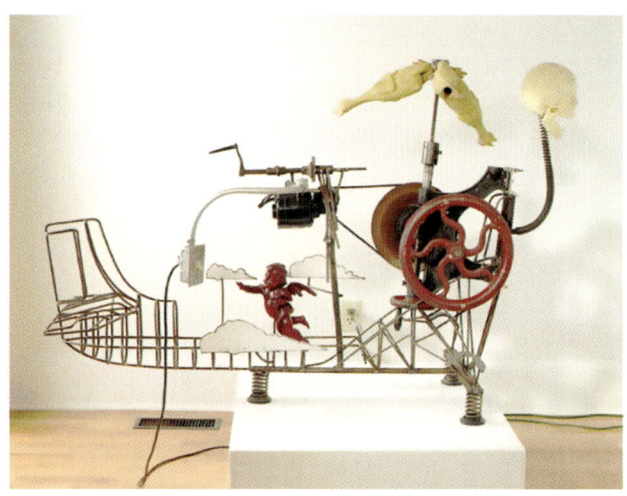

Sean Pace

Sean Pace (b. 1975) grew up in Charlotte, North Carolina, where his "hillbilly mechanic" father taught him formidable mechanical skills. The elder Pace piloted airplanes, worked on sports-car engines, and maintained a well-known biker gang's fleet of motorcycles.

Sean studied art at the University of North Carolina-Asheville, graduating in 2004. Since then he has developed a reputation for kinetic assemblages made from salvaged machine parts and industrial components. As an art student he noticed that museum and gallery visitors spent very little time looking at the art, but observed that, "once you create something that moves, they'll stare at it till sun goes down." While much of Pace's work comments upon current social issues, he sometimes treats more enduring themes, as in his kinetic piece "Death Slapper," inspired by the experience of a friend who survived a purportedly fatal illness.

Pace's "Social Landscape Painting Machine" outputs automatic drawings that reflect the bike rider's route and terrain. The painting machine part holds a roll of paper and a mechanism that squirts paint from turkey basters. By riding the vehicle through different parts of Asheville, knocking on doors, and engaging residents, Pace encourages them to participate in painting the landscape in their own neighborhoods. [Source: the artist, and author interviews with the artist]

Death Slapper, 2008, 41 x 64 x 26.5 in., post drill, electric washing machine motor, rubber chickens, plastic skull, mixed media. Courtesy of the artist.

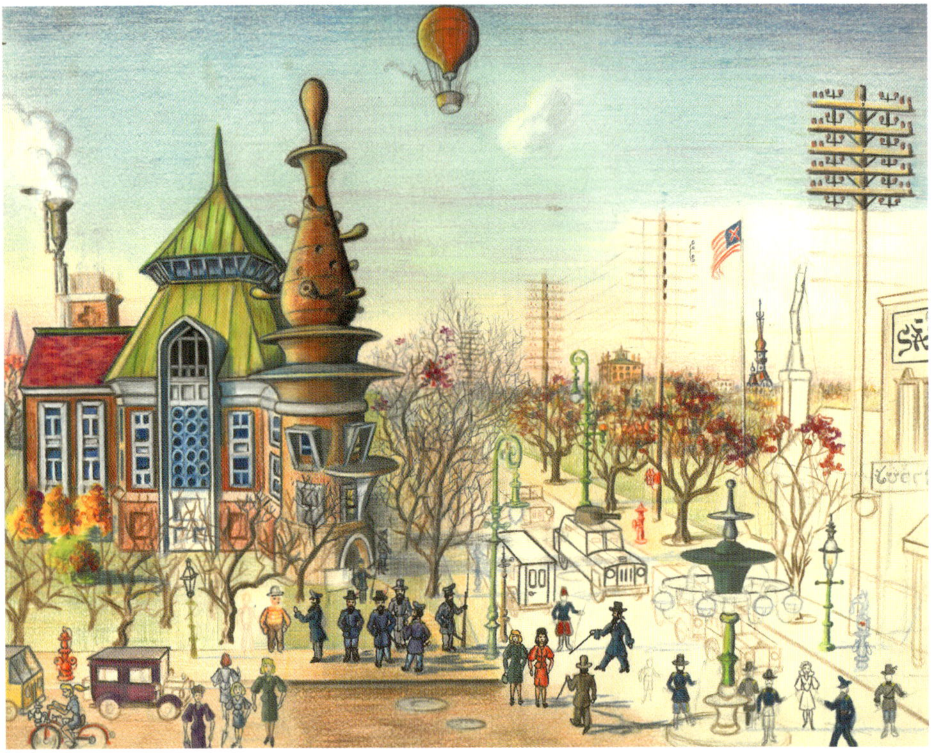

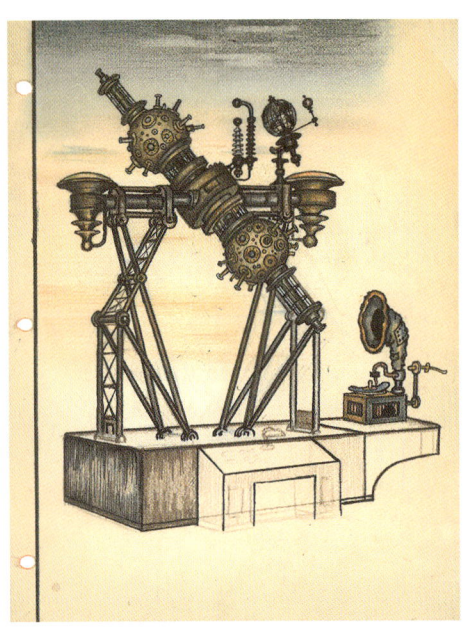

Renaldo Kuhler

Renaldo Gillette Kuhler was born in Teaneck, New Jersey, in 1931. Originally named Ronald Otto Louis Kuhler (he officially changed his name in 1967) he was the only son of Otto Kuhler, a German immigrant famed for Art Deco designs of steam locomotives. In 1948 Renaldo's father retired and moved the family from the suburbs of New York to a ranch in a remote Colorado valley .

As a way of coping with the unbearable isolation, young Renaldo imagined a world he named Rocaterrania, a tiny nation located on the border between Canada and northern New York. To help transform the fantasy into reality, he began making drawings of a vast network of imaginary friends, the city where they lived, and eventually an entire made-up country with an elaborate history that reflected events in his own life.

Kuhler recorded the story of Rocaterrania in dozens of journals and sketchbooks. Its citizens inhabited a unique form of architecture, spoke their own language, and followed their own code of ethics based on a national religion that incorporated Buddhism, Judaism, Islam and Christianity, folded in with the nature-worship of German Romanticism.

Rocaterranians dressed in their own national costumes and decorated their homes in Rocaterranian "style," which included gaslights, hand cranked phones, and riveted iron boilers. Kuhler preferred a universe where, unlike in the electronic world, mechanisms were visible and comprehensible. Because the nation had to be self-sufficient, he focused a great deal of his inventive imagination on developing sewage systems and waste-recycling plants that generated power for the country through the production of methane gas technologies that are only now becoming prevalent in the real world.

Eventually, the drafting skills that Kuhler honed in sketching his imaginary country helped him begin a career as an illustrator. After graduating from the University of Colorado at Boulder, Kuhler spent six years as exhibits curator at the Eastern Washington State Historical Society Museum in Spokane, then became chief illustrator for the North Carolina State Museum of Natural History (today's NC Museum of Natural Sciences). Now retired, Kuhler has resided in Raleigh for more than 40 years. [Source: author interviews with the artist, and Brett Ingram]

....

(opposite above) *Rocaterrania National Museum*, ca. 1960–65, 11 x 14 in., color pencil, watercolor; (below) *Untitled* (planetarium projector for museum), ca. 1960, 11 x 8.5 in., ink and color pencil. Courtesy of Brett Ingram.

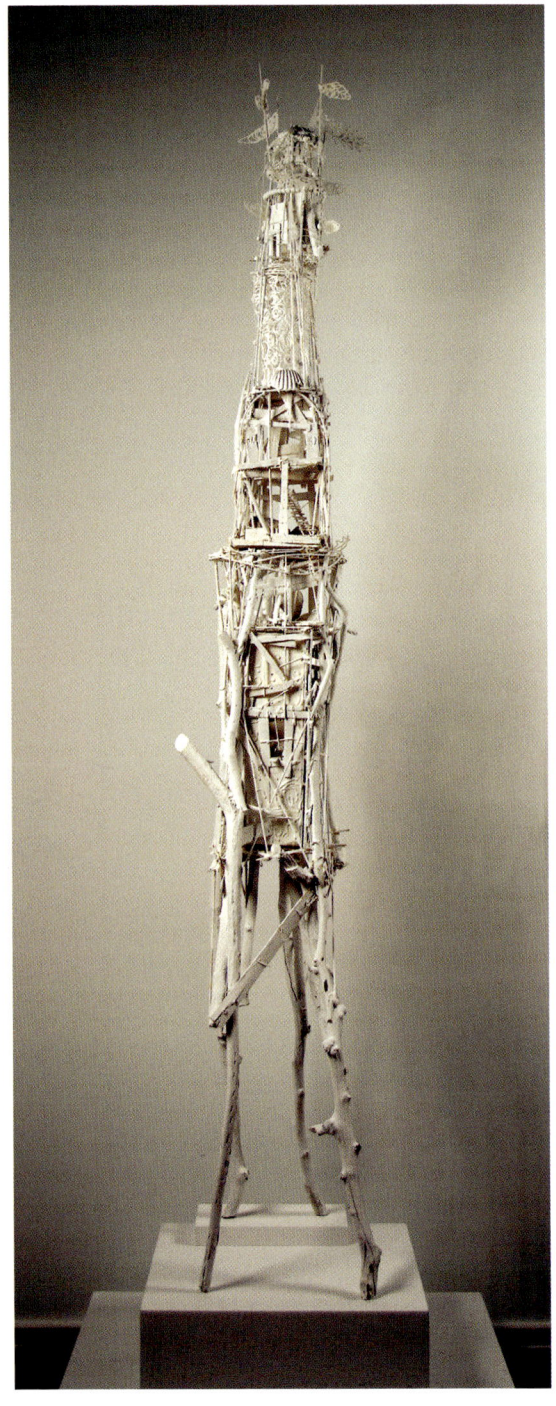

Sylvain Corentin

Sylvain Corentin was born in Montpellier, France in 1962, and currently lives in the same region. His architecturally referenced, mixed-media sculptures—called "anarchitectures"—typically begin as structures in his imagination before he starts using twigs and other found materials to build them as three-dimensional objects. He finds inspiration in geology, archeology, geography, urban planning, nature, and prehistoric monuments.

He usually finishes his tall, tapered towers by painting them white, lending them a stark, otherworldly appearance. Reminiscent of Joseph Cornell's *Dovecote* series of boxes, they suggest elaborate, fanciful, high-rise houses for birds, spirits, or a fictional tribe. Corentin's work has been featured in exhibitions in Europe since the late 1980s but wasn't shown in the United States until 2012. [Sources: Cavin-Morris Gallery, artist's website]

....

Rome Tower, 2011, 90 x 13 x 17 in., found wood, ceramics, paint.
Courtesy of Cavin-Morris Gallery, New York.

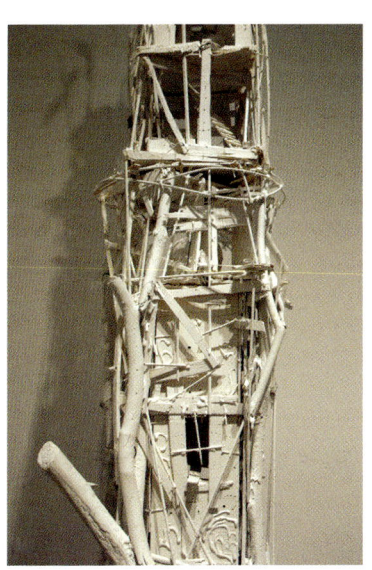
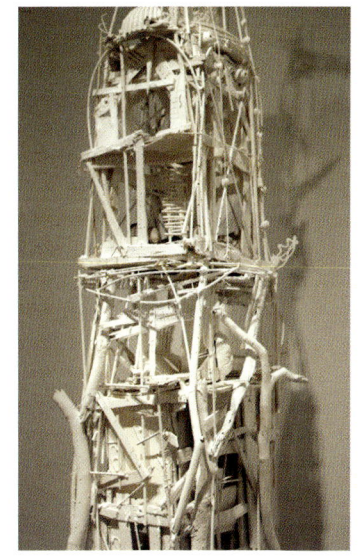

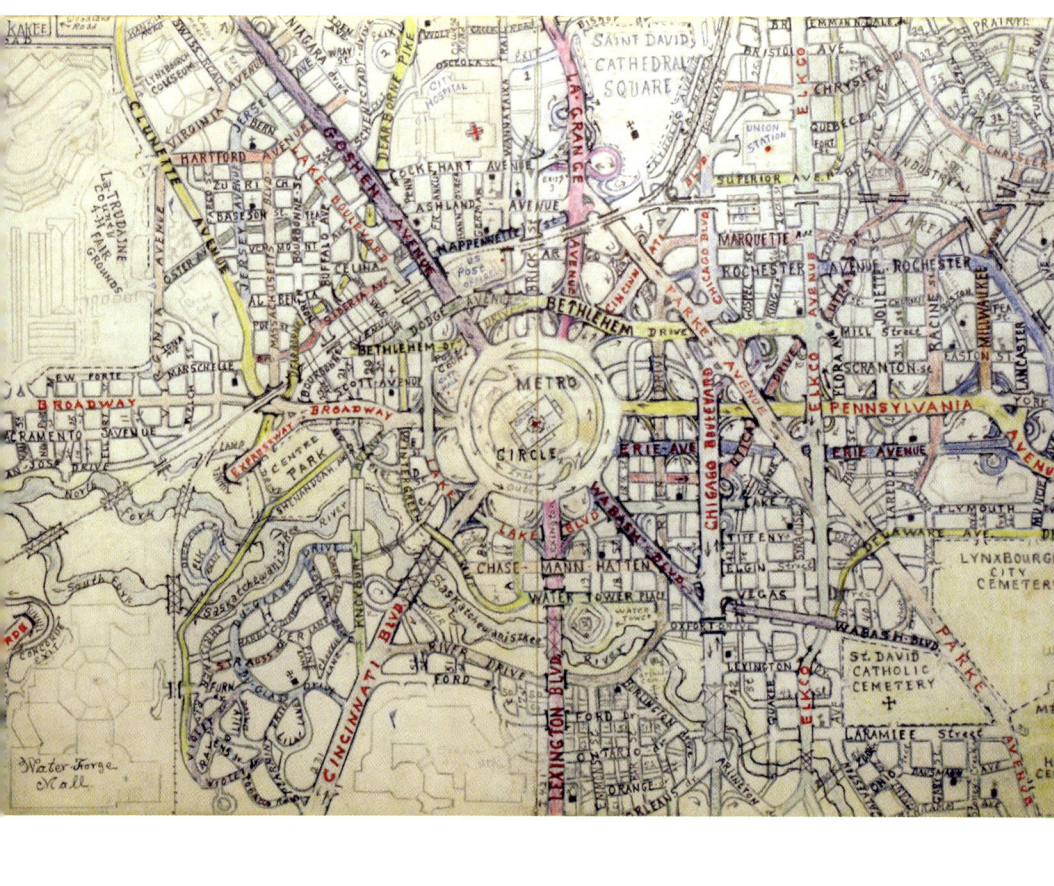

Jacques de du-Glass

James Donald Beatty was born out of wedlock and abandoned by his biological mother in South Bend, Indiana in 1931. Adopted by the Douglass family of Warsaw, Indiana, he was raised on their farm and became an expert in the genealogy of his adopted family. After discovering their French lineage, he had his own name officially changed in mid-life to Jacques de du-Glass.

Besides genealogy, du-Glass also became fascinated with the history of the homes, churches, and farms of rural Indiana, painstakingly recording them in a collection of sketchbooks. Eventually he began creating alternate histories, describing an imaginary town he named Lynxbourgh, Indiana. Du-Glass built a miniature scale model of the town in his backyard, as well.

On the back of his Lynxbourgh map, he wrote about the founding of his fantasy town, even going so far as to create an ancestor (known, as du-Glass was, as J.D.): " . . . Jacob du Glass, a man who was already well established in the family . . . was sent for and in 1847, J.D. as all knew him by, named the town on the River Linksburgh, [and] he also name the county La-Trudaine."

After a life troubled by unemployment and conflicted sexual identity, he became reclusive and devoted himself totally to making his art. In a final strange twist, after his death in 2001 he was buried in the wrong plot. [Source: Lindsay Gallery]

....

Street plan of Lynxbourgh, n.d., 12 x 18 in., ink and color pencil.
Courtesy of Lindsay Gallery, Columbus, OH.

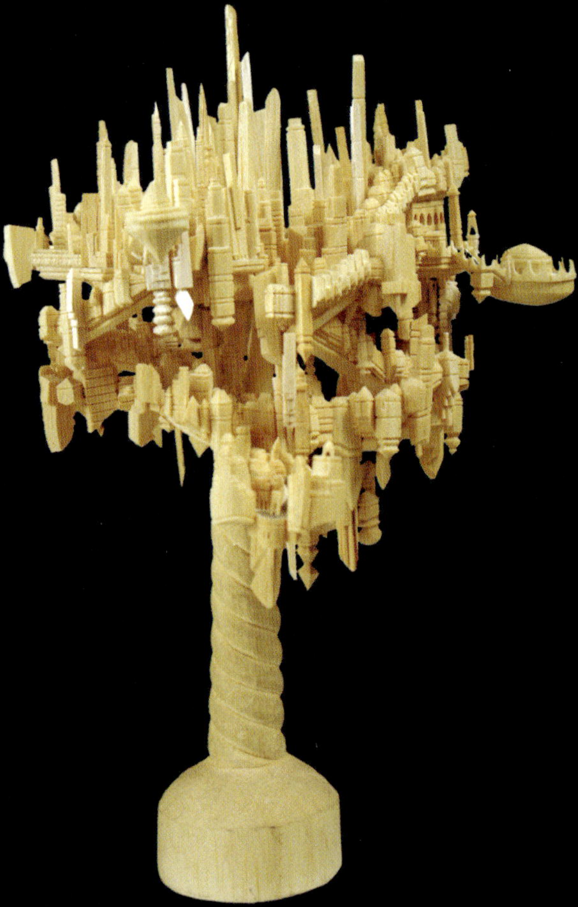

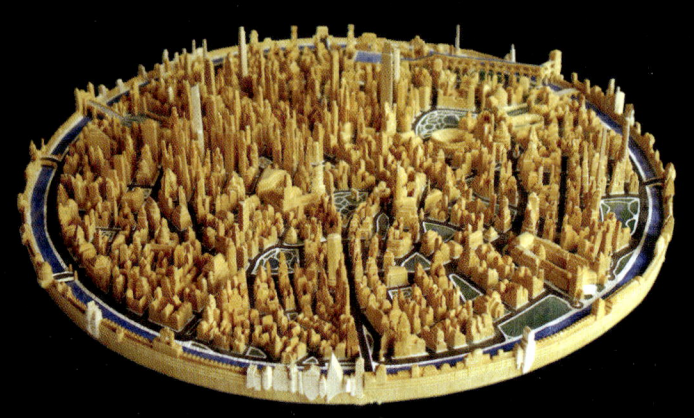

Charles Carroll

Charles Carroll was born in 1956 in Mineola, New York, and grew up in East Meadow on Long Island. His father had a small home construction and repair business, while his grandfather had been a structural engineer who had built roads, bridges, harbors, and even a gold mine in Arizona, so Charles grew up as the offspring of several generations of builders.

As a young boy he was enthralled with maps of all kinds and obsessed with cities and architecture. By age 5 he began building his own structures in the backyard, while constantly drawing floor plans and maps, both real and imaginary. At 14 he made the first of several dozen miniature cities, some of which were five feet in diameter.

His fascination with maps, architecture, and miniatures eventually led him into the museum world. After earning a degree in rhetoric from SUNY Albany in 1978, Carroll worked briefly at the Center Galleries at the State Psychiatric Center in Albany, and the Arras Gallery in New York City before being hired at the Museum of Fine Arts, Houston, in 1980. He's worked at the Fowler Museum at UCLA and the Dayton Art Institute, and is now the registrar at the Nasher Museum of Art at Duke University. About his artwork, which he makes in his spare time, he explains:

> What inspired me to start making cities on pedestals was a 1980s article in *Connoisseur* magazine about London's Museum of Mankind and its then-orphaned collection of Sapi-Portuguese ivory salt-cellars. At the time they were being ignored for being neither pure African art nor pure European art. Years later, one of my greatest thrills was working on the exhibition "Africa in the Renaissance" in Houston, which featured these very salt-cellars.
>
> I was equally interested in Chinese and Japanese scholars' stones (small imaginary worlds for one to contemplate) as well as the medieval reliquaries and carved ivories made for curiosity cabinets. Like my own street plans or maps they are also pathways for the mind and imagination.
>
> I started doing the pencil drawings in late 2000 or early 2001, after my father died. That was a difficult time for me. The drawings were almost automatic, a compulsion, with connections and reconnections of thoughts and reflections, and dead ends as well as interesting surprises and roads not taken.

[Source: the artist]

....

(opposite above) *City XXVI*, 1997–98, 13.75 x 10.5 x 8 in., balsa wood, acrylic;
(below) *City VII*, ca. 1990, 4 x 18 x 18.5 in.; balsa wood, acrylic. Courtesy of the artist.

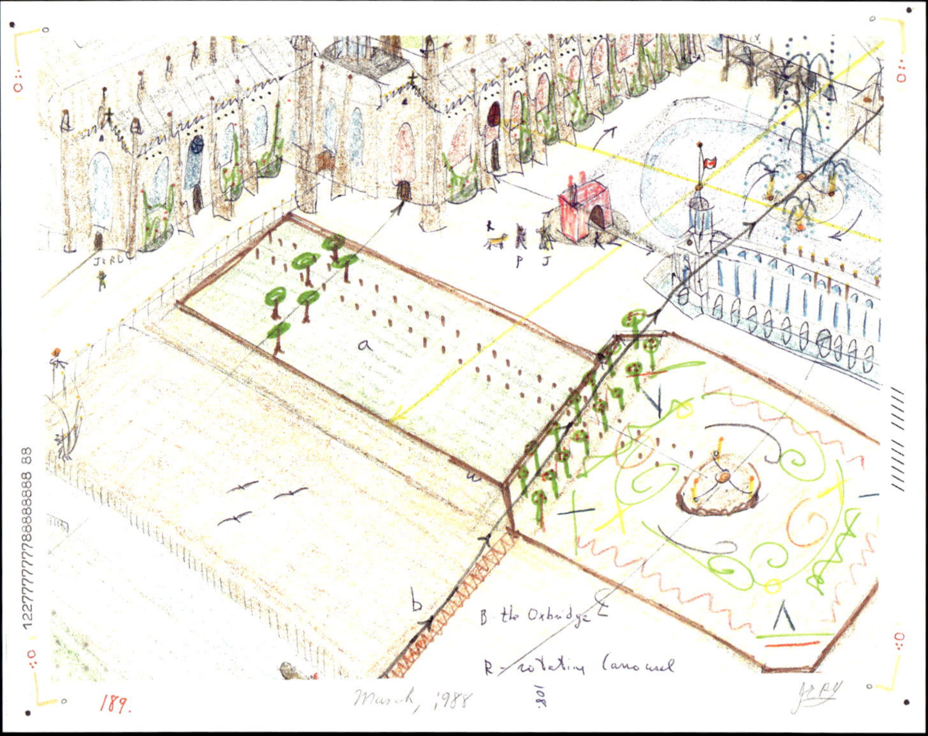

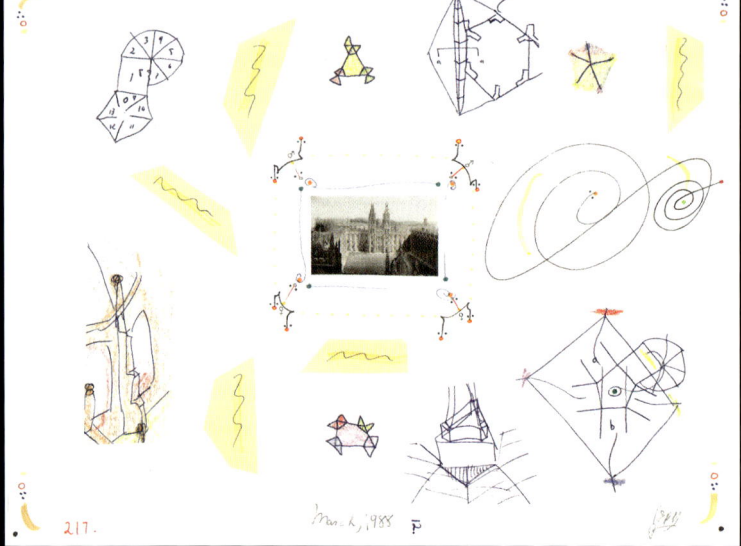

B - the Oxbridge

R - rotating Carousel

March, 1988

189.

201

217.

March, 1988

John Devlin

John Devlin (b. 1954) left his home in Nova Scotia in 1979 to study theology at St. Edmund's College of the University of Cambridge, England. He quickly found himself overwhelmed by both the beauty of the riverside setting and the historic architecture, and by the stress of a formidable academic work-load and distance from home. In Cambridge, he was "the happiest I'd ever been in my life . . . and the saddest as well."

His intense emotions triggered a manic episode, and Devlin has since been diagnosed with bipolar disorder. He returned to Nova Scotia and be-gan four years of treatment, during which he was medicated and repeatedly hospitalized.

Eventually settling in Walton, a community on the Minas Basin, Devlin started making architectural sketches on the reverse sides of forms from the local unemployment office. Working with ink, wax crayons and collage, he imaginatively transposed some of Cambridge University's oldest build-ings to the small imaginary village of Cambridge, Nova Scotia, a utopian city also partially based on Oxford University and Venice.

Over the next 15 years he made hundreds of these drawings as part of an elaborate plan for his "circular mandala city" on an artificial island. The buildings incorporate aspects of the Cambridge originals, combining fanci-ful details such as laser lights, a reflecting pool, and rotating fountains. The small symbols, colored dots and typewritten numbers in the margins of his drawings represent calculations relating to an "ideal ratio" he has derived from Cambridge's architecture. He has named his architectural fantasy *Nova Cantabrigiensis* (Latin for "New Cambridge"). It constitutes his con-cept of a place where he can regain the happiness he experienced on his best days at the university.

[Sources: Tony Thorne, "Heavenly City—John Devlin's Utopian Visions," Raw Vision #77, Winter 2012/2013 and *Nova Cantabrigiensis,* a film with Phlis McGregor, Kings College Cambridge, 2010]

....

New Cantabrigiensis drawings, 1988, all 8.5 x 11 in., color pencil, crayon, markers, collaged photo. Courtesy Henry Boxer Gallery, London.

68

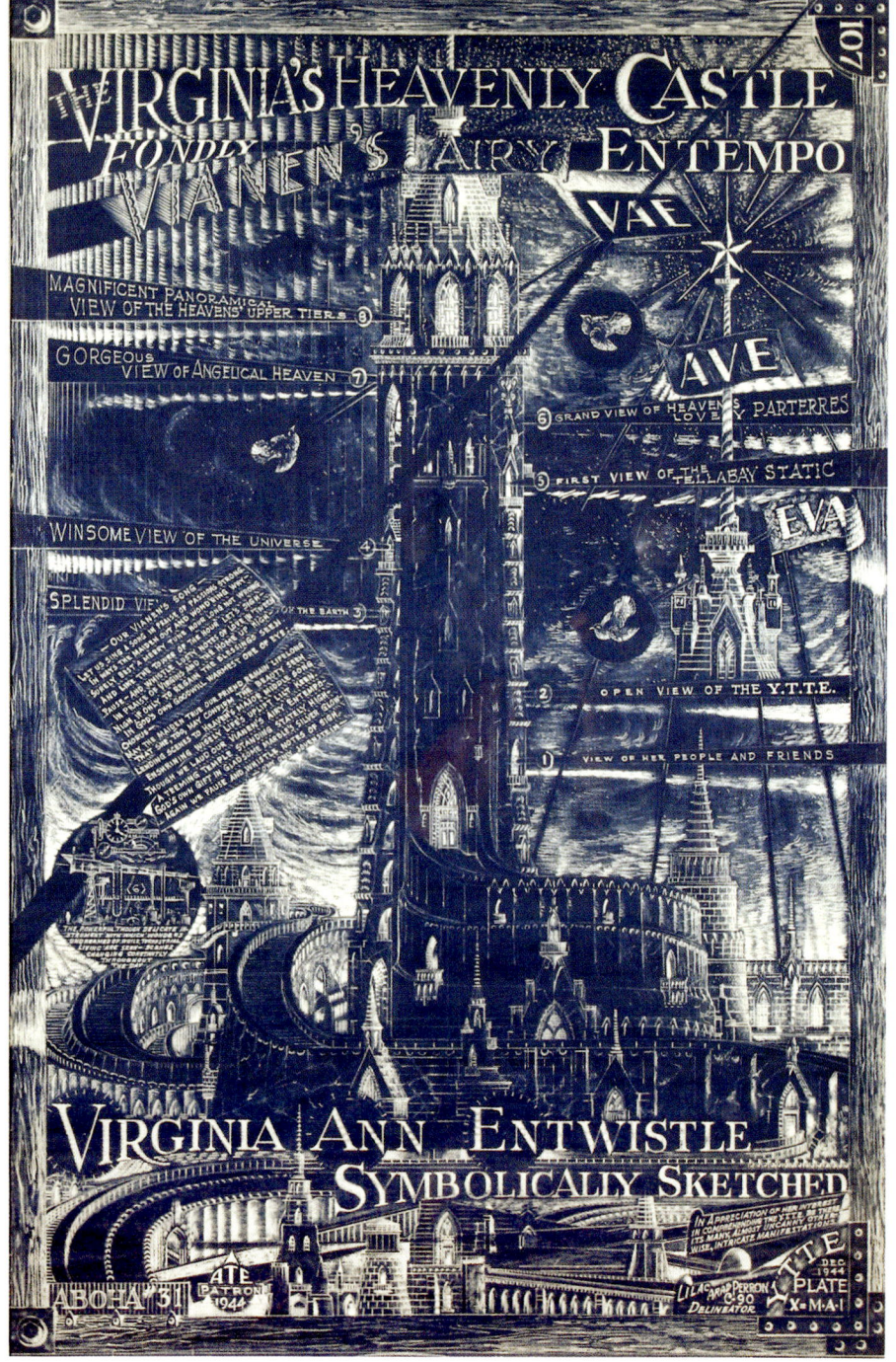

Achilles Rizzoli

In the San Francisco architecture firm where Achilles Rizzoli worked for nearly forty years, he was regarded merely as a competent draftsman. Few if any among his colleagues, neighbors, and family knew him as an imaginative, intuitive artist.

Rizzoli was born in Marin County, California in 1896, the youngest son in a family of five children, and lived most of his life in San Francisco. His father committed suicide in 1915, though his remains went undiscovered for 21 years. Rizzoli shared a home with his mother until her death in 1937, never marrying.

With little formal education, he learned drafting at a technical school and continued as a member of the San Francisco Architecture Club. Between 1935 and 1944, Rizzoli produced a body of spectacular architectural renderings in the grand Beaux Arts style. Done in colored ink on rag paper, many of the drawings were intended as "symbolic portraits" of friends and family, depicting them as buildings. Five birthday tributes to his mother, which he called "the Kathredals," rank among his most elaborate architectural portraits.

Other drawings represent plans for a fantasy "expeau" inspired by the Panama Pacific International Exposition, which was held in San Francisco in 1915 to mark the completion of the Panama Canal the previous year. Rizzoli called his imaginary fair the "Y.T.T.E.," which stood for "Yield To Total Elation." The major units or buildings in the Y.T.T.E. comprise another series of drawings.

After an unproductive hiatus, Rizzoli began a new project in 1958, filling large sheets of vellum with poetry, prose, architectural drawings, quotations, commemorations of events (church burnings, or the death of JFK), or some combination of these. Together, the 350 sheets of vellum comprise the A.C.E. ("AMTE's Celestial Extravaganza"). Rizzoli made up his own language, replete with many symbolically loaded puns, anagrams and solecisms ("earchitecturally," for example, or "Architecture Made To Entertain"—hence, AMTE).

In 1977, while working on a sheet of the A.C.E. entitled *Rest in Peace Awhile*, Rizzoli suffered a debilitating stroke and spent the last five years of his life in a nursing facility. He died in 1982. The drawings and writings he left behind document a life lived, in his own words, "in an unbelievably hermetically sealed spherical inalienable maze of light and sound, seeing imagery expand in every direction." [Source: the Ames Gallery]

....

Virginia's Heavenly Castle [or] Virginia Ann Entwistle Symbolically Sketched, 1944, 35.5 x 23.5 in., cyanotype blueprint. Courtesy of the Ames Gallery, Berkeley, CA.

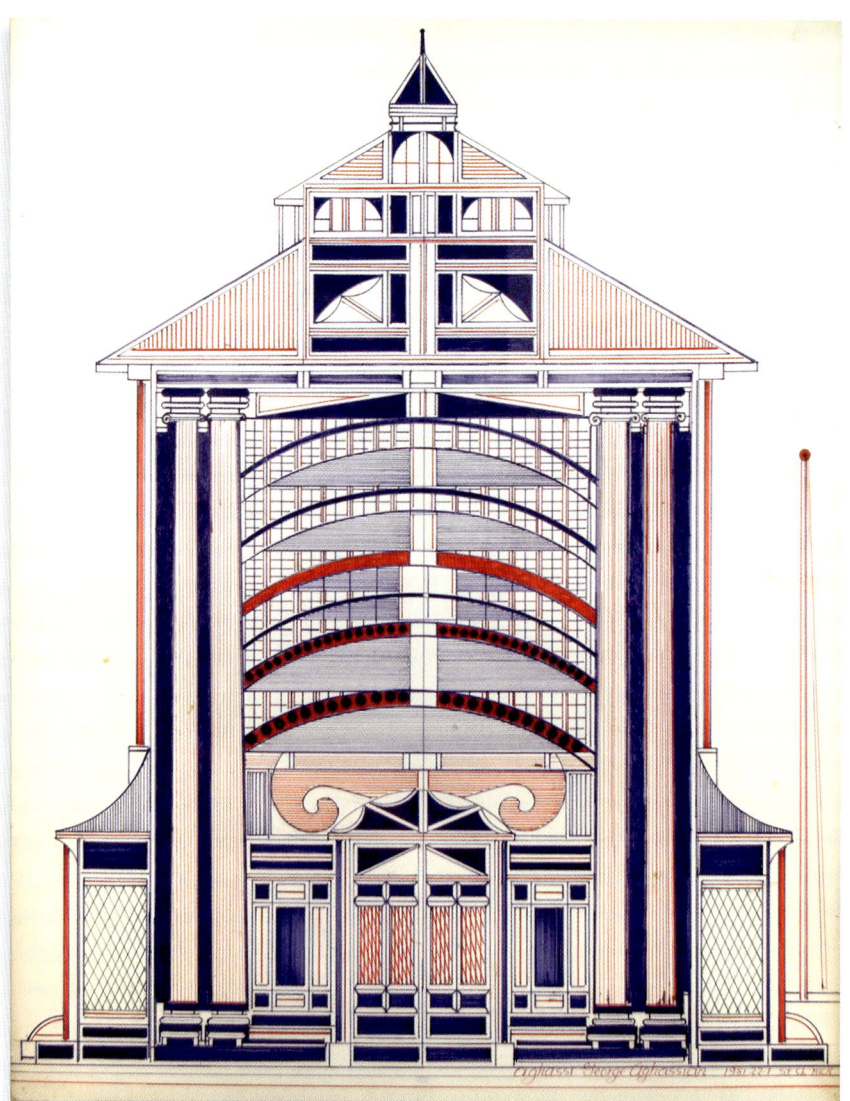

George Aghassian

Remembered by neighbors as quarrelsome and litigious, George Aghassian wrote rambling letters to President Gerald Ford and drew private architectures using only pens, a straight edge and a French curve. The structures he depicted in these compositions seem both ancient and modern—almost believable but, at the same time, fantastic. The strange images that emerge from the patterns he drew can be viewed simultaneously as buildings and abstract forms.

Aghassi George Aghassian was born in Turkish Armenia around 1904, and came to America in 1921. Intending to study architecture, he struggled through the Great Depression, working as a ship's cabin boy and as a waiter in New York City.

He and his wife eventually opened a restaurant in Memphis, Michigan in 1965, where he took up drawing while recovering from an illness. Although his earliest drawings date to 1972, his most complex works were made between 1979 and 1983. When he died in 1985, nearly 100 plans for buildings were found among his effects, revealing that he never abandoned his dream of becoming an architect. [Source: Lindsay Gallery]

....

Untitled (building in St. Cloud, Michigan), 1981, 28 x 22 in., ink and marker.
Courtesy of Lindsay Gallery, Columbus, OH.

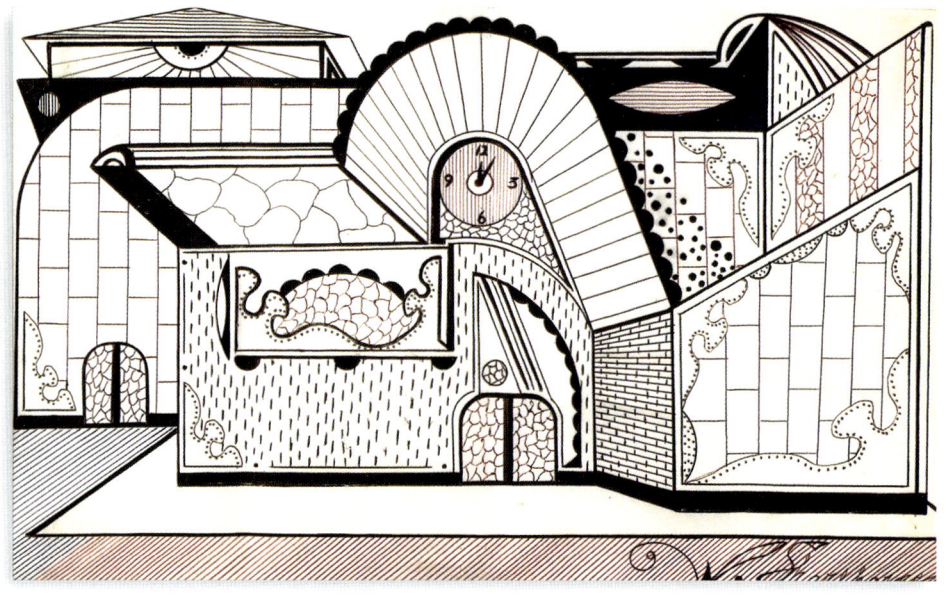

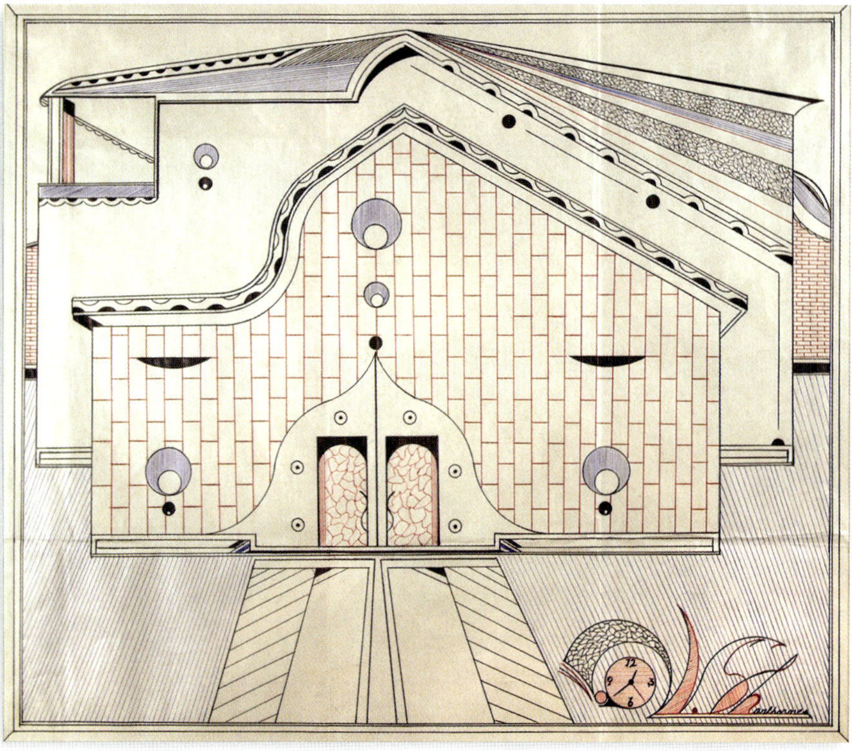

Welmon Sharlhorne

Welmon Sharlhorne was born in Houma, Louisiana, in 1956, and has spent most of his life in New Orleans. He began making architectural line drawings during one of several prison sentences for extortion and drug-related violations totaling about 25 years.

After his release from the Louisiana State Penitentiary at Angola in 1992, Sharlhorne showed a sheaf of about 40 drawings to a New Orleans art dealer. His ballpoint images feature a graphic style reliant on precise, hard-edged lines drawn using a six-inch ruler and round container lids as guides. Some of his buildings are reminiscent of prisons, while others tend toward the mysterious and exotic.

The clock faces on many of these buildings reflect a prisoner's preoccupation with time's passage, and Sharlhorne's fondness for drawing buses like those used in interstate transit indicates a similar fascination with free movement. The dragons that appear in some of his works can be viewed as symbolic representations of his own inner "monsters" or obstacles he has faced in his struggle to make his way in a society seemingly unsympathetic to his problems.

Some of his close associates have blamed a fondness for casino gambling for his occasional homelessness but, despite a tendency to lose any money that comes his way, Sharlhorne always manages to be dapperly dressed whenever he appears in public. [Sources: Marcia Weber, Anton Haardt]

....

(opposite above) *Untitled*, n.d., 8.5 x 14 in., ink;
(below) *Untitled*, n.d., 19 x 22.25 in., color ink on gray paper.
Courtesy of American Primitive Gallery, New York.

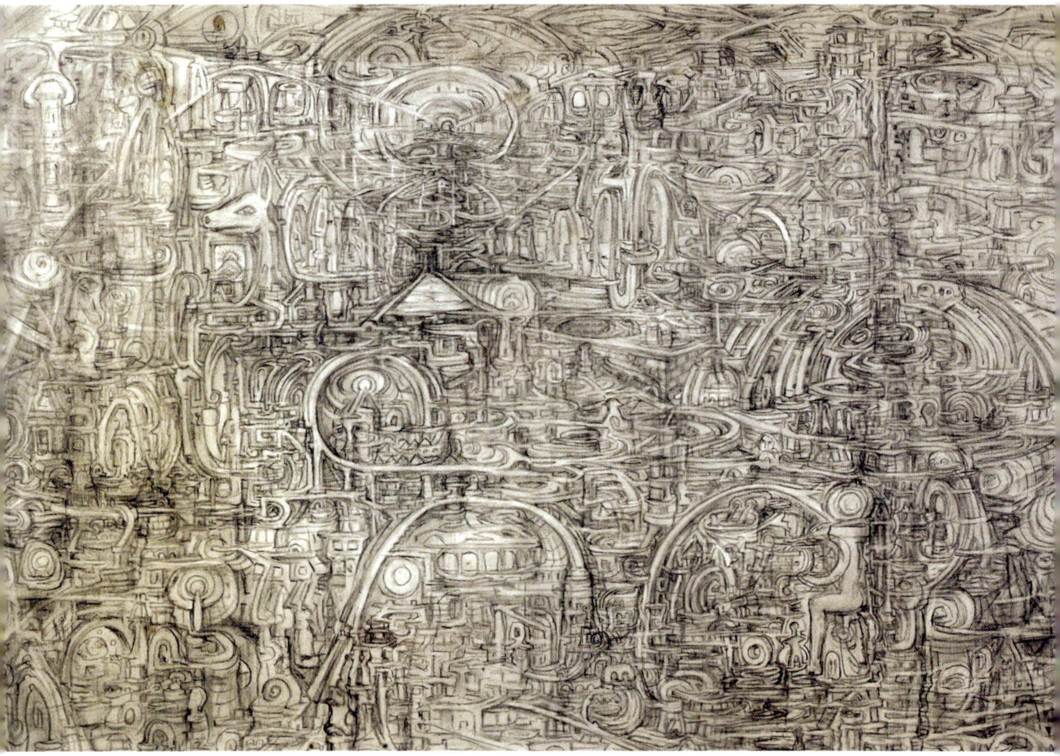

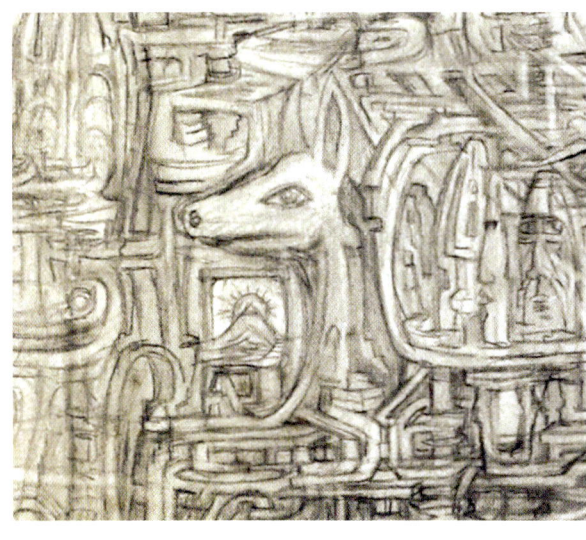

Henry Hill

Henry Hill was born in Oak Park, Illinois, in 1918, where his father and grandfather had been furniture manufacturers. Hill studied sculpture at Cornell but dropped out in 1939 claiming to have "majored in [playing] bridge and weekend parties." After failing to find work with the fledgling Disney Company in California, he joined the Army just as the U.S. entered the Second World War.

As a First Lieutenant, Hill fought in North Africa and Italy, but cracked under the strain of constant battle. He spent much of the latter part of the war as a PTSD casualty in the psychiatric ward of an Allied military hospital, wracked with guilt that other members of his unit hadn't survived combat.

After the war, he moved back to Los Angeles where he studied art on a disability pension at the University of Southern California. He became involved in the Dianetics movement and also became intrigued with Reichian therapy, a radical form of psychotherapy. After attending California Chiropractic College, Hill earned a living as a chiropractor and Reichian therapist until the mid-1960s, when, in his late 40s, he fell into the hippie drug culture that was surging through the West Coast.

Around this time, Hill began treating patients with LSD and colored light therapies, while beginning to paint obsessively in response to his own experiments with LSD and other psychotropic drugs.

Hill also ran for political office several times—once as a Republican, but in favor of legalizing marijuana, prostitution and gambling. Never elected, he subsisted on a meager military disability pension while continuing to cram his tiny South Hollywood apartment with paintings and drawings. He was in his late 70s before he dared to show them to anyone else. Of himself he sometimes said that the fictional character he most identified with was Pinocchio, "the wooden puppet who always wanted to live as a whole man." [Source: author interviews with the artist]

....

Molecular Concourse / Rampart and Rafter, ca. 1960, 20 x 32 in., graphite pencil. Anonymous lender. Detail below.

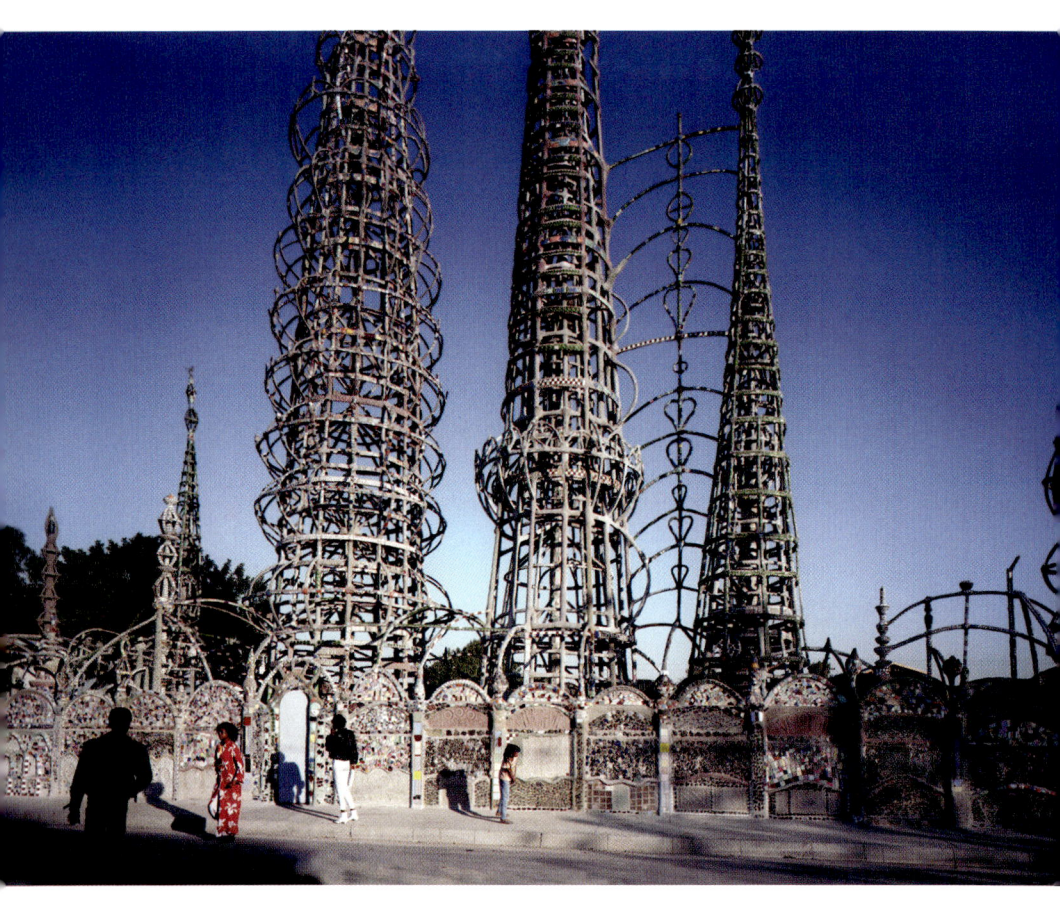

Ted Degener, photograph of Sabato Rodia's *Watts Towers*, 1987,
27.75 x 39.25 in., archival inkjet print. Courtesy of the artist.

Sabato Rodia

Sabato Rodia was born in a small village east of Naples, Italy, in 1879. As an adult living in Watts, a lower-income suburb of Los Angeles, California, he was known to friends and neighbors as "Sam Rodilla." On historical markers and in the National Register of Historic Places he is now officially (if incorrectly) known as "Simon Rodia" while his environmental creation is called "The Watts Towers."

As a young man, Rodia joined his older brother in the Pennsylvania coal mines but when his brother died in an accident, Rodia moved to California, settling in Watts in 1920. He purchased a triangle-shaped empty lot located at 1765 East 107th Street and began to construct a cluster of 17 interconnected towers and other structures now regarded as a major world architectural treasure. Calling them Nuestro Pueblo ("Our Town"), Rodia built the towers entirely in his spare time while working as a telephone-line repairman, tiler, and part-time security guard.

For 34 years, Rodia worked alone without powered machinery, scaffolding, welding equipment, or even basic hardware like bolts or rivets. Lacking drawings or plans, he used only common hand tools, a pair of pipe fitter's pliers, and a window-washer's safety belt.

The largest of the towers is just six inches short of 100 feet tall and contains the longest reinforced concrete column in the world. Like all the other components, it is constructed of salvaged re-bar and steel rods, wrapped in wire and coated with mortar embedded with blue and green broken glass, sea shells, tiles, and other found objects, much of it contributed by local children. Faucet handles, horseshoes, corncobs and handprints pressed into the wet mortar added decorative indentations. To shape the metal bars into the complex loops and curves he needed for the towers, Rodia bent them cold by inserting them under some nearby railroad tracks and levering them by hand a little at a time.

In 1954, when Rodia was approaching 75, he suddenly stopped, gave his property to a neighbor, and moved to Martinez, California, to be nearer to his surviving family. He ceased making any kind of art, and although he showed up long enough to be applauded at a 1961 conference on the towers held at the University of California, Berkeley, he never returned to the site or saw the towers again.

In 1956 a fire destroyed the little house where he had lived while working on the site. The city Department of Building Safety summarily declared that the towers were in danger of collapsing, and ordered that they be de-

THE WATTS TOWERS

Sabato Rodia's heroes were Copernicus, Galileo, and Columbus, and he spoke of his work as celebrating their spirit of exploration. He also told interviewers that he started working on his project to keep himself busy after he quit drinking.

The triangular shape of the lot, created by the diagonal slash of railroad tracks running alongside his property, was ideal for the display he created. At the narrowest point, Rodia built "Marco Polo's Ship," a bench that resembles a four-tiered galley with seashell-encrusted masts. Behind this are three lofty towers—including the 99-foot giant—with four smaller companions, ranging from ten to fifty feet high. The towers create the outline of the composition and ensure that "Nuestro Pueblo" is visible for several blocks.

The base of the triangle, closest to the site of Rodia's, contains park-like elements: a gazebo-arbor, stalagmite groupings, fountains, birdbaths, and benches. The basic materials of the towers are steel rods surrounded by concrete over chicken wire. The towers' lacy use of space comes from the interlocking circular tiers of vertical columns, reinforced with woven spiral and elliptical horizontal rings and spokes connecting the individual towers to create delicate, lace-like shapes soaring into the sky. The mosaics embedded in the mortar waterproof the cement.

The Watts Towers, featured in the 1961 Museum of Modern Art exhibition, "The Art of Assemblage," are now recognized as one of the finest examples of American environmental art. Representing the power of creative vision, the towers uplift the Watts community while serving as an urban oasis and providing a dignified public space for ceremonies. They've also become the site of the Watts Towers Arts Center, which offers classes in the arts to the community and sponsors several annual music festivals.

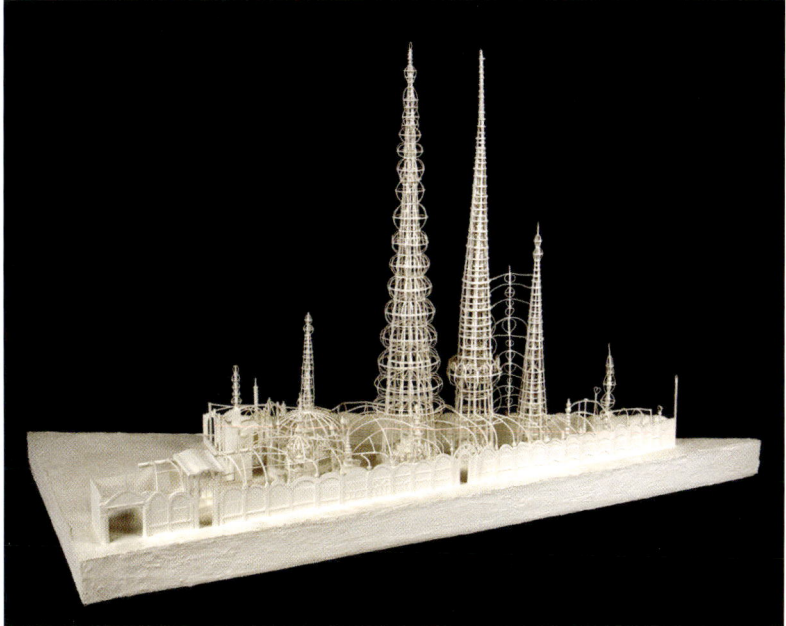

molished and the lot cleared. But a group of concerned citizens, including architects, artists and museum curators, insisted on an engineering test to assess the Towers' real strength and safety. In October 1959 steel cables were attached to each tower so a crane could exert lateral force. The towers didn't budge and the test was stopped when the crane itself began to fail.

In 1978, the state assumed ownership of the property and designated it the Watts Towers of Simon Rodia State Historic Park. The towers became a National Historic Landmark in 1990. As a cultural icon, the Watts Towers have been recognized in all kinds of features, ranging from the BBC television series *The Ascent of Man* to an episode of *The Simpsons*. At the same time, Simon Rodia earned a place of honor in American art history, not least for being included next to Bob Dylan on the original 1967 cover of the Beatles' landmark *Sgt. Pepper* album. Unfortunately Rodia himself did not live long enough to enjoy his worldwide fame, dying in 1965. [Sources: Seymour Rosen, S.P.A.C.E.S., and Watts Towers Art Center]

....

Larry Harris, model of Sabato Rodia's *Watts Towers*, 1997,
27.5 x 49 x 19.5 in., painted and glued matboard, wood.
Courtesy of the artist.

TOWERS OF BABEL

The necessities of life—food, clothing, and shelter—have been the focus of rituals since time immemorial. The saying of blessings before meals, like the sacraments performed in special costumes worn at christenings, graduations, weddings, and funerals, has long given food and clothing special meaning. Shelter has always included a religious or mystical aspect, too. Formalities like the laying of cornerstones, the rites of builders' guilds (the secrecy of Freemasonry is only one example), ceremonial dedications, and all the folklore regarding tools or domestic features like ladders, keys, thresholds and hearths, underscore the sacred value we place on homes, schools (ask any alum), workplaces, and houses of worship.

The latter are especially important and that importance has often been symbolized by their sheer height. The taller a building is, the more sacred and majestic it feels. Towers, church steeples, temples, pyramids, pagodas, minarets, cathedrals, and ziggurats express, through their elevation above the surrounding landscape, the age-old desire to extend the human reach high into the spiritual realm.

Fringe builders, no less than their architecturally trained counterparts, have felt this same desire for extended stature. Their ideas have occasionally been realized (the Watts Towers or Howard Finster's multi-tiered polygonal church in Pennville, Georgia, are two good examples), but most of their imaginative output has been limited to drawings or models.

Professional architects sometimes experience the same frustrations. In 1956, Frank Lloyd Wright proposed a mile-high skyscraper for downtown Chicago that would have had 528 stories and included 184,600,000 cubic feet of space. If it had been built, the "Illinois Sky-City," or "The Illinois," would have been twice as tall as the current tallest building in the world, the Burj Khalifa in Dubai. So far, however, engineering and materials science have not advanced enough to enable a tower as tall as the one Wright proposed to safely support its own weight.

Wright might have been considered a crank or madman if The Illinois were the only building he ever proposed, or if none of his structures had ever been realized. Sabato Rodia, meanwhile, made no proposals at all but simply picked up his tools and began building the Watts Towers, and succeeded.

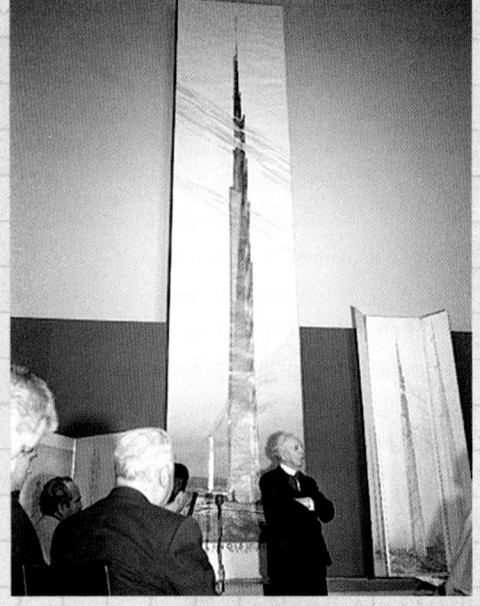

Frank Lloyd Wright at press conference, October 16, 1956, courtesy of *Chicago Sun-Times* Archives

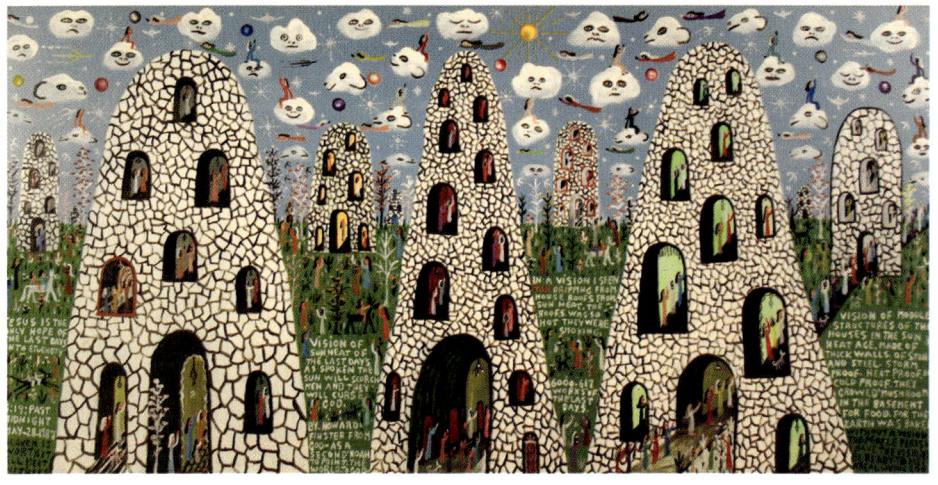

(above) Howard Finster, *Vision of Moddle Structures*, 1987, (Finster number 6000 and 617), 23.75 x 47.25 in., enamel on panel. Courtesy of John Denton.

(right) Howard Finster, *Heavenly Mansions*, n.d., 16 x 11.75 in., color marker and pencil. Courtesy of Jane and Bert Hunecke.

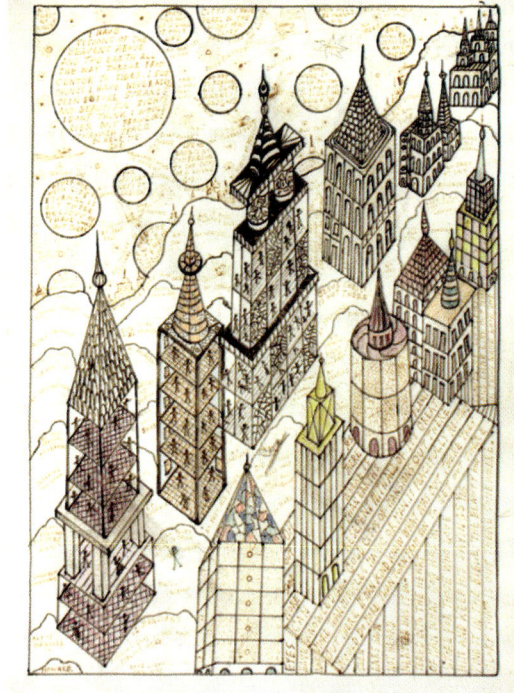

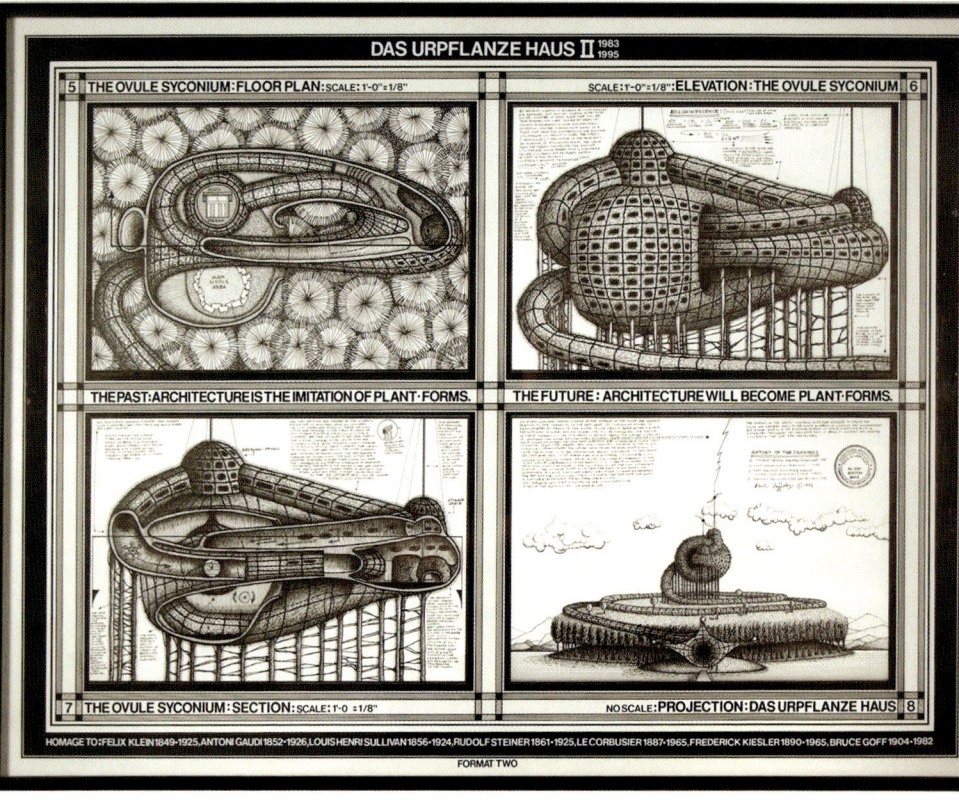

DAS URPFLANZE HAUS II 1983 1995

| 5 | THE OVULE SYCONIUM: FLOOR PLAN: SCALE: 1'-0"±1/8" | SCALE: 1'-0"±1/8": ELEVATION: THE OVULE SYCONIUM | 6 |

THE PAST: ARCHITECTURE IS THE IMITATION OF PLANT·FORMS.   THE FUTURE: ARCHITECTURE WILL BECOME PLANT·FORMS.

| 7 | THE OVULE SYCONIUM: SECTION: SCALE: 1'-0 ±1/8" | NO SCALE: PROJECTION: DAS URPFLANZE HAUS | 8 |

HOMAGE TO: FELIX KLEIN 1849-1925, ANTONI GAUDI 1852-1926, LOUIS HENRI SULLIVAN 1856-1924, RUDOLF STEINER 1861-1925, LE CORBUSIER 1887-1965, FREDERICK KIESLER 1890-1965, BRUCE GOFF 1904-1982

FORMAT TWO

Paul Laffoley

Paul Laffoley was born into an Irish Catholic family in Cambridge, Massachusetts in 1940. He spoke his first word, "Constantinople," at six months, then remained silent until the age of four (having been diagnosed as slightly autistic), when he began to draw and paint. In his senior year at Brown University, he underwent eight electric shock treatments. He was dismissed from the Harvard Graduate School of Design and apprenticed with the sculptor Mirko Baseldella before going back to New York to work with visionary architect Frederick Kiesler.

In 1968, Laffoley returned to Boston and moved into an 18- by 30-foot utility room to found a one-man "think-tank" and creative unit he called the Boston Visionary Cell (BVC). Laffoley supported himself with a job at the Boston Museum of Science, returning to the BVC not only to eat and sleep, but also to work on multimedia renderings of his visions of alternative futures and complex realities.

During a routine CAT-scan of his head in 1992, a miniature metallic implant was discovered in the occipital lobe of his brain. Local members of the Mutual Unidentified Flying Object Network (MUFON) declared it to be an alien nanotechnological laboratory. Laffoley has come to believe that the implant is extraterrestrial in origin and is the main motivation behind his ideas and theories.

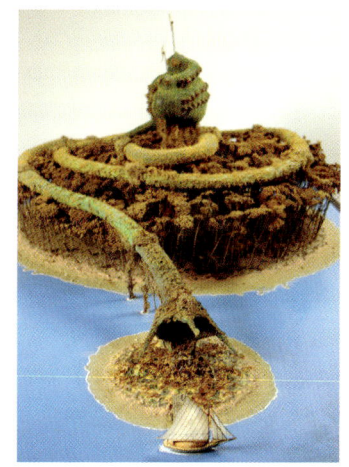

Laffoley has proposed that his *Urpflanze Haus*, a Klein-Bottle-shaped structure to be grown by genetically modifying tree DNA, may offer a way to construct bases on other planets, where delivering building materials to construct shelters may prove extremely difficult. It might be considerably more efficient for a robot to plant a packet of seeds designed to thrive in, say, the Martian environment, and transform local soil into something like wood (if not grow the entire shelter), than to send spaceship-loads of raw materials in advance of a human landing. [Sources: the artist, and the Cartin Collection]

....

(opposite) *Das Urpflanze Haus II (5–8)*, 1993–95, 27 x 36 in., drafting ink and presstype. Courtesy of the Cartin Collection.

(above) *Model of Das Urpflanze Haus*, 1997, 25.5 x 44.5 x 63.75 in., mixed media. Courtesy of the artist and Kent Fine Art, New York.

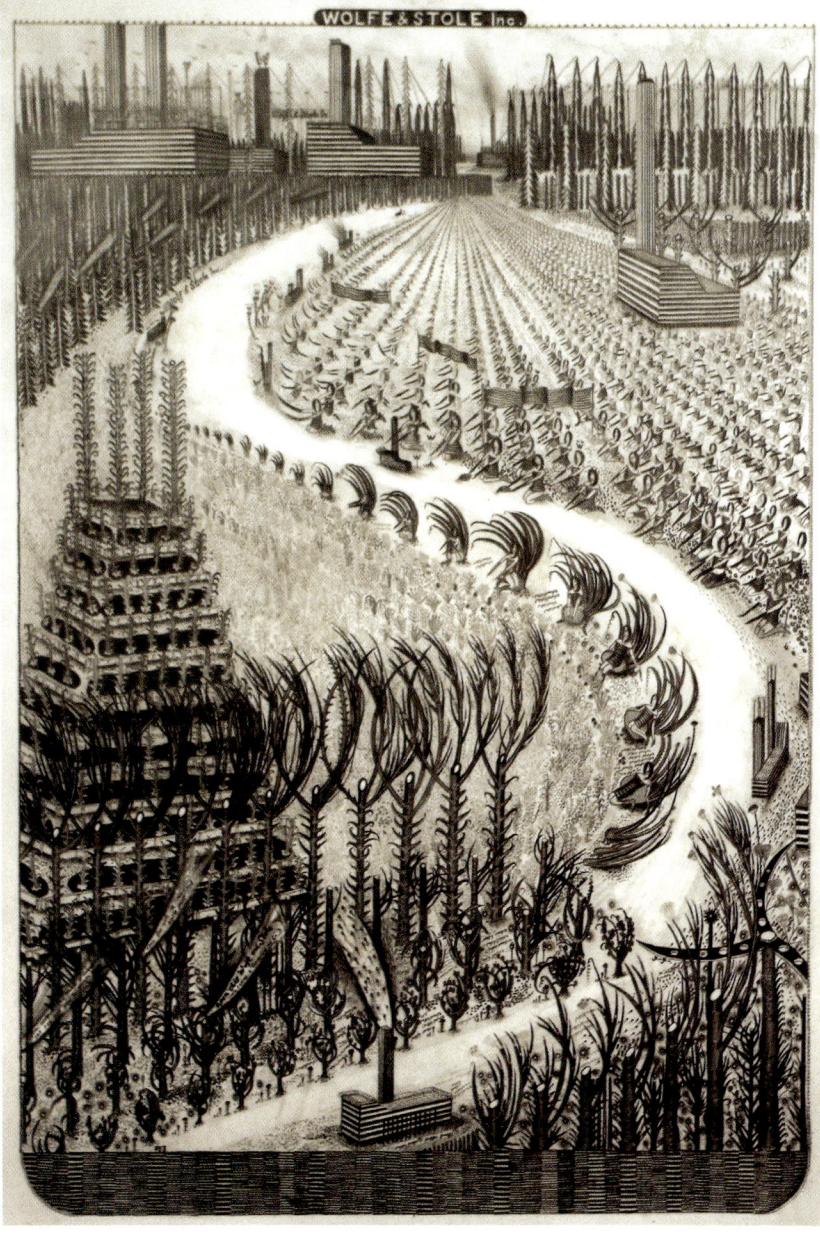

Chris Hipkiss

Born in 1964, Chris Hipkiss grew up in a middle-class western suburb of London. He left school at 16 to apprentice as a cabinetmaker at his father's joinery firm. Soon thereafter he began to draw. At age 21, Hipkiss married a computer analyst, and together they are now raising a family in a small village in the English countryside.

Or so the story goes. In fact, "Chris Hipkiss" is a fictional creation of Alpha and Chris Mason, two artists living in the French countryside who have collaborated for more than 30 years. The couple, who refer to themselves as "Team Hipkiss," has written that "Hipkiss is not a male artist in essence; Hipkiss is an entity that wouldn't exist, as such, without the feminine partnership that created it." The Hipkiss universe they have created is drawn with "hypnotic precision and restraint, intricately repetitive and laying bare an anthropomorphic, post-industrial world populated by mutant cyber-dominas—an army of Hipkiss alter-egos."

[Source: Cavin-Morris Gallery and Galerie Suzanne Zander]

....

Wolfe & Stole, Inc., n.d., 33.25 x 23.5 in., graphite pencil.
Detail below. Courtesy of Cavin-Morris Gallery, New York.

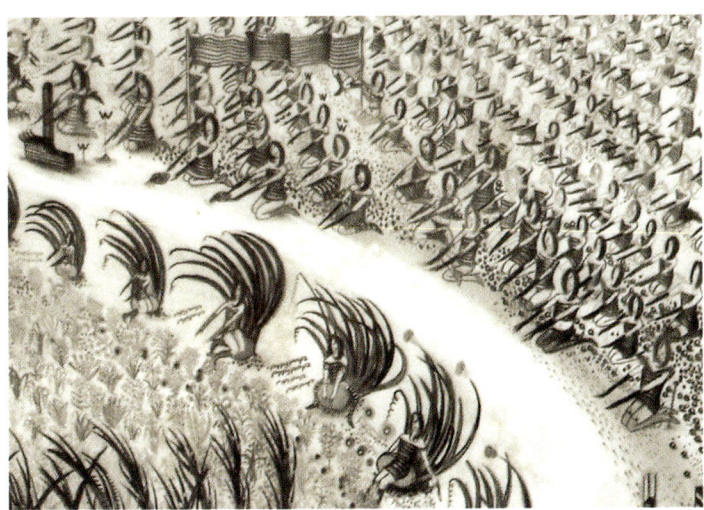

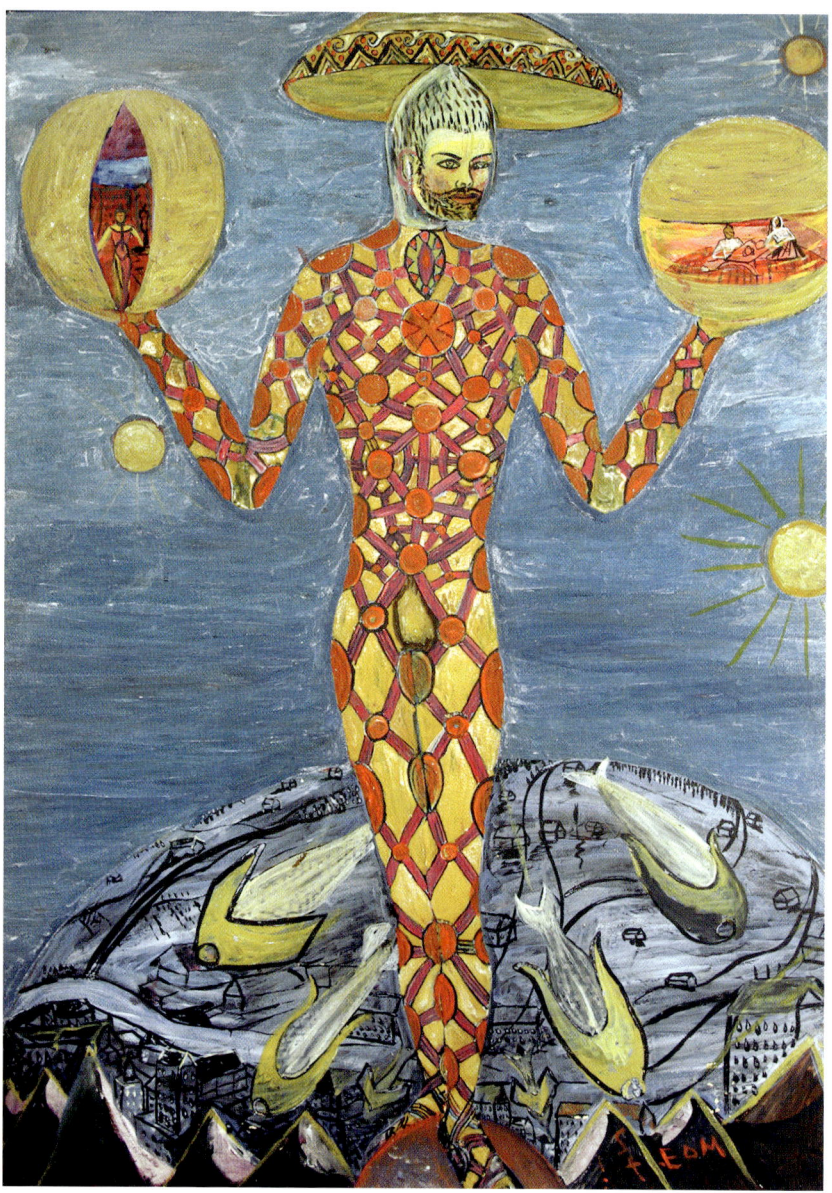

Eddie Owens Martin (aka Saint EOM)

Eddie Owens Martin (a.k.a. Saint EOM) was born in 1908 into a family of poor white sharecroppers in Marion County, Georgia. At 14 he fled the farm and hitchhiked to New York, where he fell in with an underworld community of street hustlers, drag queens and drug dealers.

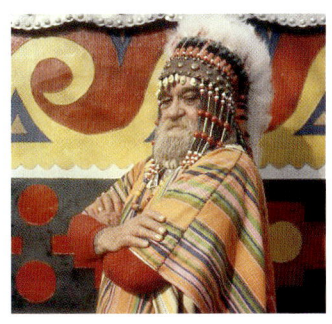

In his early thirties Martin emerged from a period of illness and personal crisis because of a visionary experience. A giant figure appeared and admonished him to change his life and become "Saint EOM the Pasaquoyan." As Saint EOM, Martin immediately began to visually articulate an alternative world around this new identity. He grew his hair and beard long and combed them up into a knot at the top of his head like a Sikh. Using brightly colored fabrics and bold patterns reminiscent of those he'd seen in anthropological films and museum displays, he began fashioning his own clothing.

At the same time he began making countless drawings and paintings based on his dreams and waking visions. These included ritualistic scenes and figures clad in form-fitting, air-conditioned suits designed for levitation. During this transformative period Martin told fortunes in tearooms along Manhattan's 42nd Street, while making art in his small Midtown apartment in the evenings.

Martin returned to Georgia in the 1950s, inheriting his late mother's farmhouse and four acres in the region where he grew up. Earning a living as a psychic, he transformed the property into a visionary environment known as "Pasaquan," now overseen by a non-profit preservation group and open to the public. Saint EOM died from a self-inflicted gunshot in 1986, following a series of debilitating medical problems. [Source: author interviews with the artist]

Pasaquoyan Man in Space in the Future, ca. 1950, 29.5 x 21.5 in., oil and metallic enamel on panel. Courtesy of the Pasaquan Preservation Society, Inc., Buena Vista, GA.

(above) Photograph of Martin as St. EOM by Jonathan Williams, 1982.

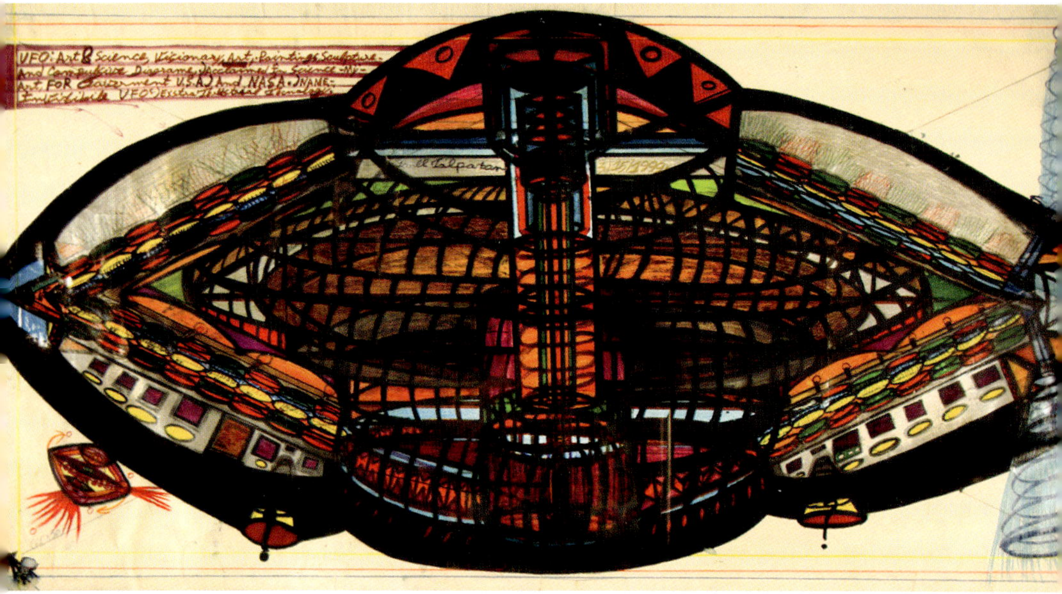

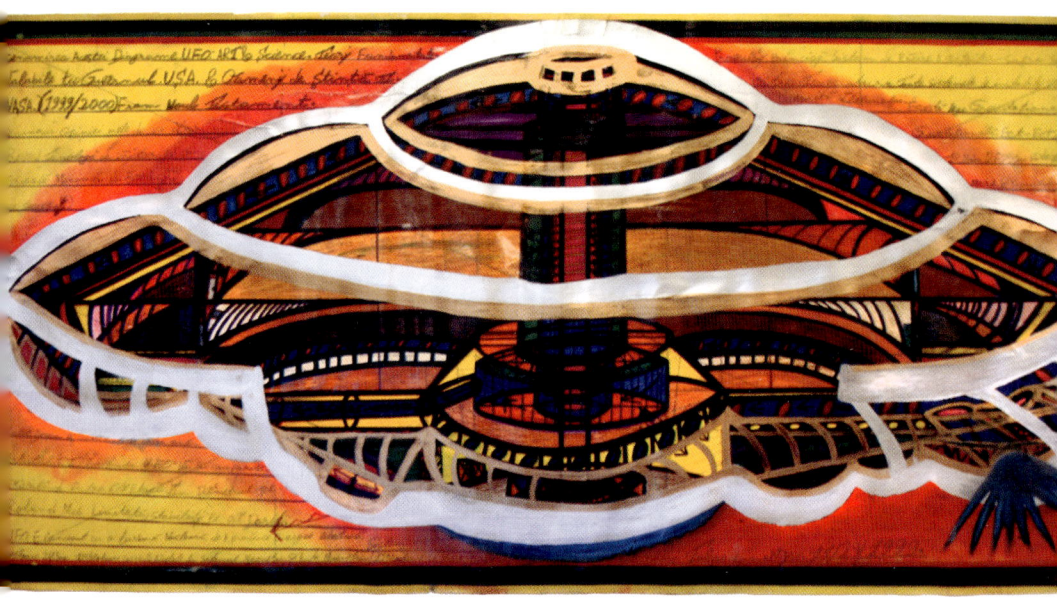

Ionel Talpazan

Ionel Talpazan was born prematurely in Petrekioaia, Romania, in 1955, shortly after his twin brother died in their mother's womb. Given up by his unmarried birth parents at six, he was adopted by abusive foster parents in the rural village of Miineasca. When he was seven he fled their home one night to escape a beating and, while hiding, found himself bathed in blue light given off by a disk-shaped craft hovering overhead. Henceforth, he compulsively talked about UFOs and made artworks based on the spacecraft.

One night in 1987 he swam the Danube River to Belgrade in order to escape Romania's Communist government. He found his way to the United States as a political refugee and has lived in New York since the late 1980s.

Talpazan creates drawings, paintings and carved-plaster sculptures of UFOs in his small Harlem apartment. He sells work on the street and through several galleries. Larger, more elaborately colored and detailed images reveal the internal workings of UFOs and are accompanied by explanatory texts in Romanian or phonetic English. Talpazan has said that his ultimate goal is to educate his fellow earthlings about the mysterious technology and hidden meanings of UFOs, and he remains hopeful that his efforts will eventually draw the attention of NASA. [Source: Daniel Wojcik, "Ionel Talpazan's Mysterious Technology," in *Raw Vision* #45, Winter 2003/4]

(opposite above) *UFO*, 1999, 30 x 61 in., poster paint, pencil, ink, oil crayon on paper. Courtesy of Henry Boxer Gallery, London.

(below) *UFO*, 1999, 30 x 63 in., poster paint, metallic paint, pencil, ink, oil crayon. Courtesy of Jack Stevens.

IT'S A BIRD! IT'S A PLANE!

While UFOs are often associated with 1950s sci-fi films and TV shows, sightings have been documented as long as history has been recorded. An Egyptian papyrus from the reign of Thutmose III (1504 to 1450 BC) records that "in the year 22, of the 3rd month of winter, sixth hour of the day [there] was a circle of fire that was coming in the sky . . . . It had no head [and] the breath of its mouth had a foul odor. Its body [was] one rod long and one rod wide . . . . After some days passed, these things became more numerous in the sky than ever. They shone more in the sky than the brightness of the sun . . . . [After dark], these fire circles ascended higher in the sky towards the south . . . "

On July 5, 592 BC, the Hebrew prophet Ezekiel looked up at the sky and saw a mysterious "wheel in the middle of a wheel" piloted by "four living creatures" and described it in the first chapter of the biblical *Book of Ezekiel*. Though often interpreted as a prophetic vision, the exact date and attention to details (how the wings were attached, how the wheels retracted when the vehicle lifted off, the sound it made as it departed) have led many researchers to wonder whether Ezekiel may actually have encountered a UFO.

The Roman historian Livy (Titus Livius Patavinus) also mentions sightings of shining, fiery objects shaped "like round shields" at Arpi and Capua in 216 BC, in Tarquinia in 99 BC, and near Spoletium in 90 BC. But the earliest use of the term "flying saucer" appears in a Japanese document that refers to an object observed on the night of October 27, 1180, as looking like a flying "earthenware vessel." Meanwhile William of Newburgh chronicles another late 12th-century sighting that terrorized the monks of Byland Abbey in North Yorkshire, England, which was shaped like a silvery "discus."

On April 8, 1897, hundreds of people in Wilmington, North Carolina, saw an airship described as having many colored lights and a brilliant searchlight that shone over the city. It came in from over the ocean, moving west at high speed. Although possibly an early dirigible (Zeppelin prototypes date to 1896), its high speed and the fact that no attempt was ever made to publicize or profit from the event makes this seem unlikely. Dozens of other reports of sightings in Europe and the U.S. in 1896–97 prompted H.G. Wells to write *The War of the Worlds*, published in 1898.

Flying saucers captured the inventive imaginations of homespun cranks and tinkerers for other reasons, many of them undoubtedly psychological. A vehicle that could be piloted by a single person and that could theoretically lift off from any backyard or rooftop without requiring a long runway or even a helicopter's need to avoid tree branches or power lines, and that could zip off on a tangent in any direction at the operator's whim, must seem like the quintessence of freedom.

By contrast, the hassle of other forms of air transport—making reservations, buying a ticket, sitting in an assigned seat—or automobile travel—staying on the road, following predetermined routes, obeying traffic laws—make them seem like exercises in lockstep conformism. The rash of flying saucer sightings that characterized the late 1940s and early 50s probably owed as much to the wistful dreams of newly discharged GI's facing the conventionality of returning to civilian life as to the disconcerting scientific breakthroughs of the fledgling atomic age.

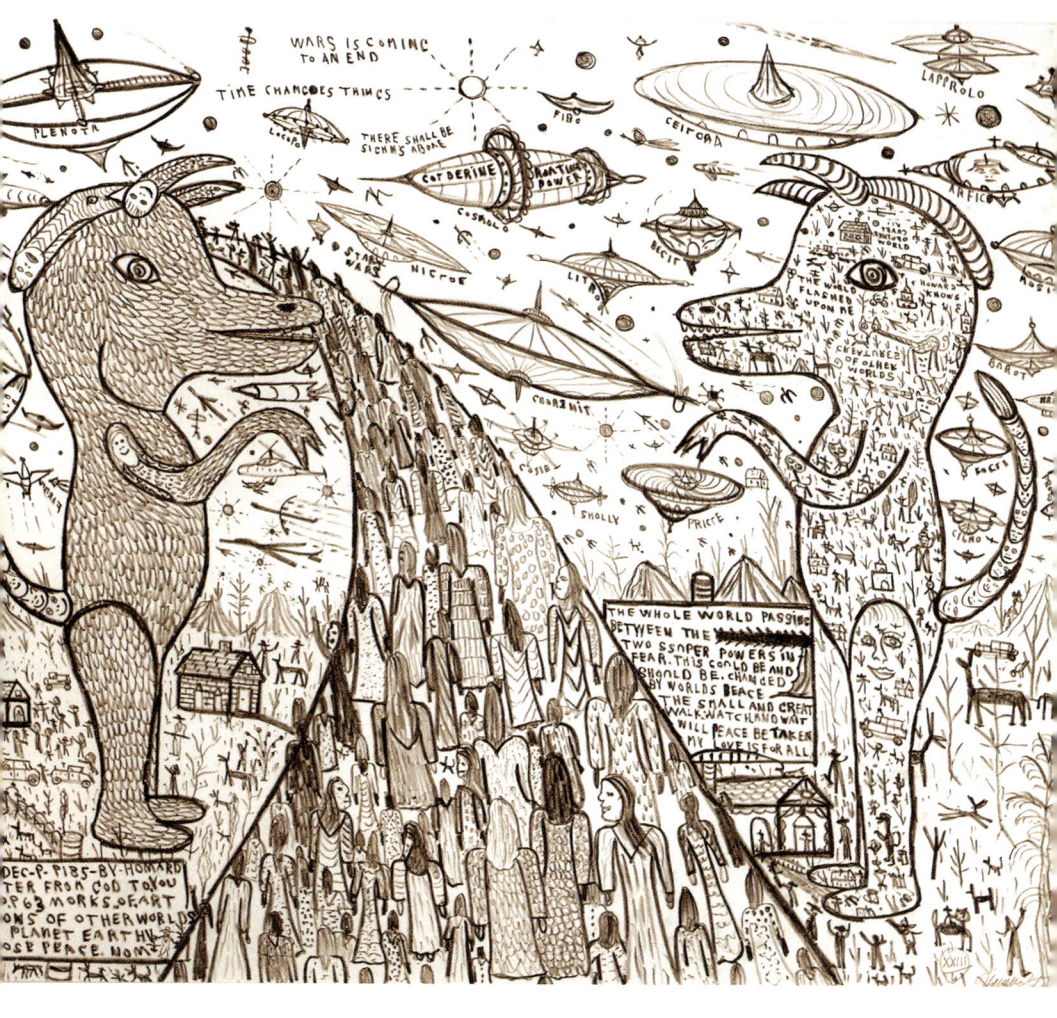

Howard Finster, *The Whole World Passing Between the Two Superpowers in Fear*, 1985 (Finster number 4000 and 963), 21.5 x 26.75 in., monotone lithograph. Anonymous lender.

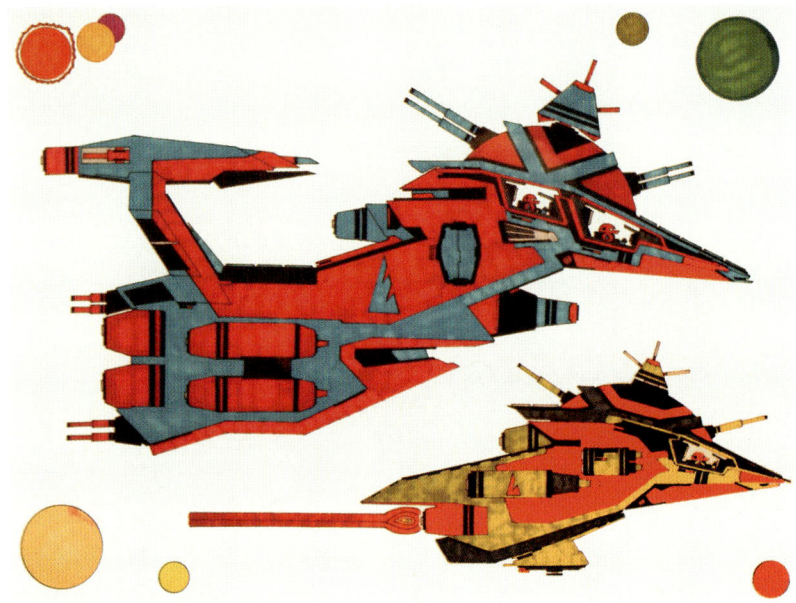

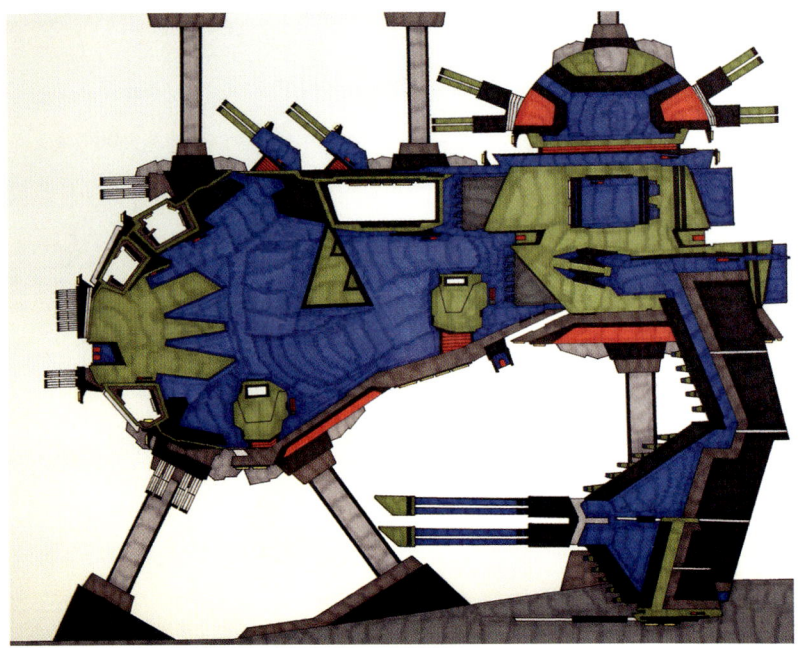

James Wong

James Wong is a self-taught artist born in Hong Kong in 1972. He now supports himself in New York City as a dishwasher and laborer. Working with colored-ink markers, Wong focuses exclusively on space exploration technology and defensively-armed, futuristic aircraft in his drawings. His images are inspired by science-fiction films and TV shows, comic-art graphics and video games. [Source: American Primitive Gallery]

....

(opposite above) *Untitled*, 2009, 17.5 x 23.5 in.,
(below) *Upgrade Hanger 79B*, n.d., 18.5 x 23.5 in.,
both ink and marker on Bristol paper. Courtesy of
American Primitive Gallery, New York.

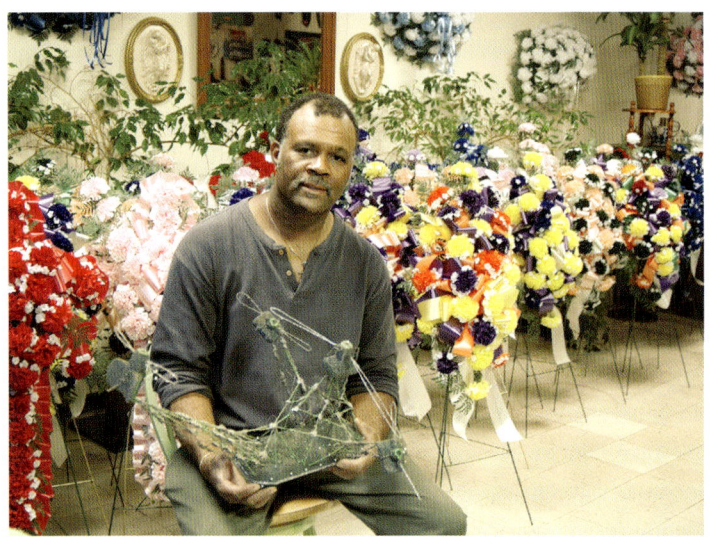

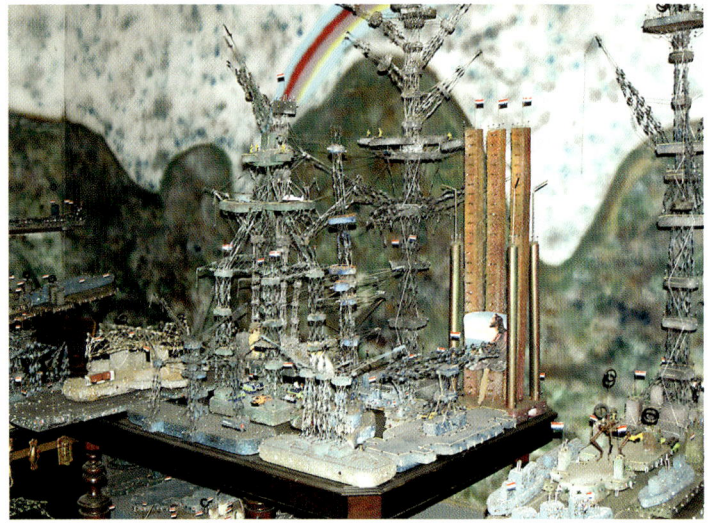

Richard Brown

Richard Brown runs Brown's Flower Shop in Littleton, NC, where he makes floral arrangements for proms, funerals and weddings. The front of his shop displays artificial bouquets that show the range of his skills, as well as a personal collection of country antiques and artifacts swapped with customers who couldn't afford to pay for their flowers with cash.

But behind the counter, a door leads to a private workroom filled with visionary sculptures made from Styrofoam, floral wire, hot glue, and glitter, depicting alien spacecraft, skyscrapers and military vessels. Brown refers to the entirety of it as "the Future Past." Suspended from the ceiling are airplanes, helicopters, and a wide rage of UFOs. Below them, cities with heliports and cranes are surrounded by barges and aircraft carriers that continue Brown's militaristic theme. It's all constructed with the leftovers from funerary displays.

He was inspired, he says, by a near-death experience in 1980 in which alcohol poisoning put him into a 48-hour coma. He passed through a dark tunnel into the bright light of revelation, but it wasn't until his mother was hospitalized in 1995 that Brown started making the models in his spare time to distract himself from obsessing about her illness. "She's the cause to this . . . that's why I want to name them all Louise," he said when asked how he titled "Tower Louise 2000."

Brown feels that his work is a collaboration with Jesus, who took him forward into the future and allowed him to look back at a past that is yet to come. Although Brown was never in the military, he says, "I feel like I've been through a war . . . and my sculpture shows the future past." Because of this, all the futuristic models have a timeworn patina from their "past" military applications, but as projected from a viewpoint two to six hundred years into the future. [Sources: Everett Adelman, Lynch Collection of Outsider Art at NC Wesleyan College, and author interviews with the artist]

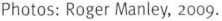

(opposite above) Richard Brown with *Helicopter*, 2009.
(below) One corner of Brown's *Future Past*
environment, 2012. The three towers are 38 in. tall.
(right) *Helipad* 11.5 x 17.5 x 13.5 in., florist foam, wire,
faux pearl pins, glitter. Courtesy of the artist.
Photos: Roger Manley, 2009.

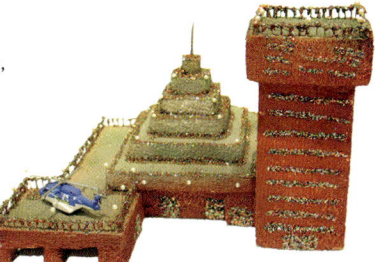

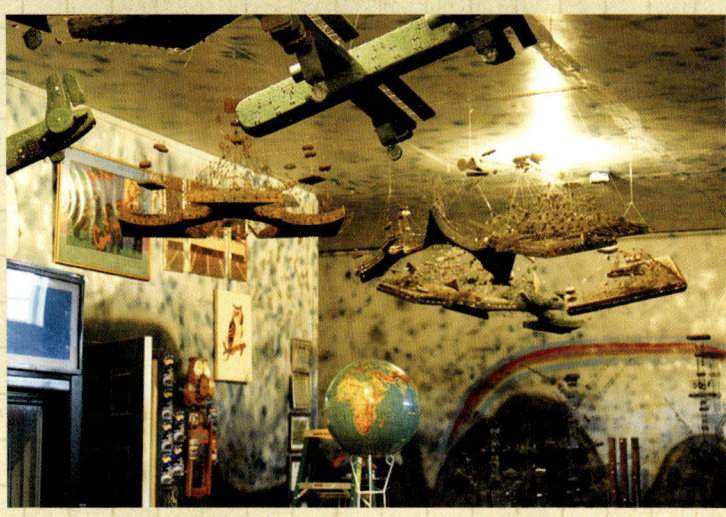

DEEP SPACE CONSTRUCTION

Richard Brown's sculptures are extremely lightweight and delicate, but surprisingly sturdy, since the internal "cellular" structure of the hardened foam lends rigidity to the structure. The flight bones of birds' wings are similarly porous for the very same reason.

This may suggest some practical ideas for real future space vehicles. Instead of trying to defy gravity by lifting entire spacecraft from the earth's surface into orbit, or hauling heavy metal components into space to be assembled there, could expanding foams make it possible to mold spacecraft in space, where aerodynamic constraints are practically nil?

Some foams are "self-healing," too, which means that leaks or potential damage from micro-meteorites might become less of a threat than they are to metal- or ceramic-skinned vehicles. Linking components with flexible wires or rods that could be delivered in long coils might prove more functional in the weightlessness of space than using stiff struts and beams that need to be delivered entire. Could the future of space exploration lie in a more improvised, less pre-planned approach?

....

Richard Brown, spaceships from *The Future Past* as installed in the workroom of his Littleton, NC, flower shop. Florist foam, wire, faux pearl pins, glitter.

Paul Esparza

One of eleven children in a Mexican-American family, Paul Esparza was born in 1963 and raised in the Midwest. Though developmentally disabled since birth, he is the only member of his family to have shown any artistic ability. He lives at home with his parents and works as a janitor at a nearby hamburger restaurant.

Esparza is fascinated by futurism, transportation, and popular culture. Many of his paintings are of public transport systems or of commercial settings such as bowling alleys, supermarkets, and RV parks. Esparza abstracts these everyday settings into futuristic fantasies. [Source: the Pardee Collection]

....

Trains & More, 1995–96, 9.5 x 12.75 in., acrylic on canvas panel.
Courtesy of the Pardee Collection, Iowa City, IA.

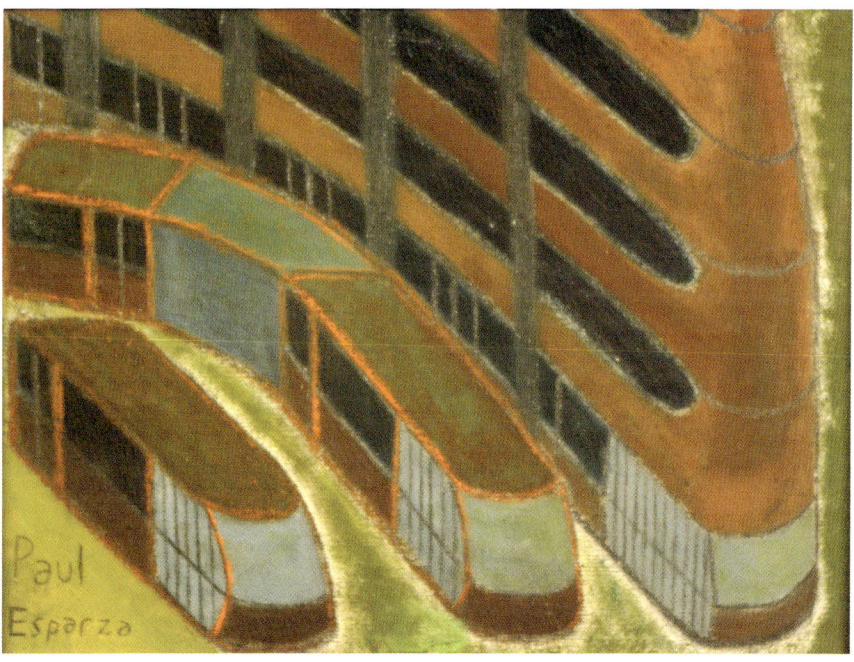

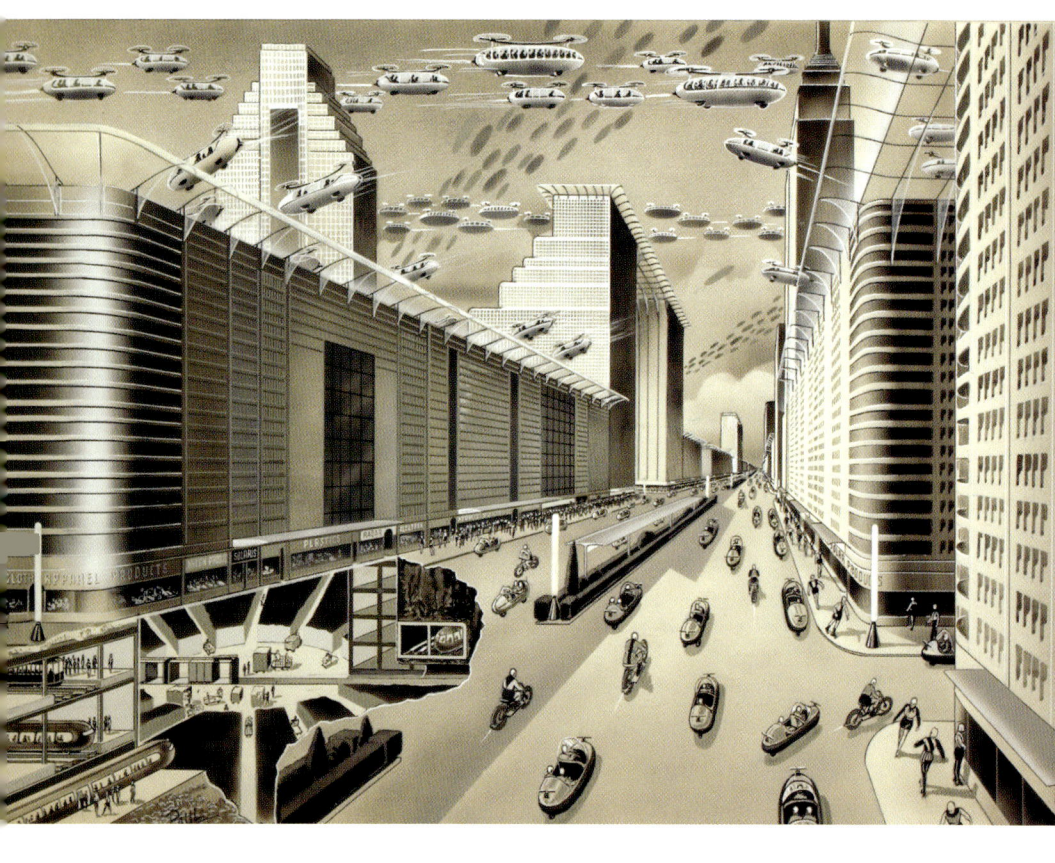

Frank R. Paul

Frank Rudolph Paul was born in Vienna, Austria, in 1884. Despite his parents' hopes, he lost interest in a career in the clergy and studied art in Vienna and Paris instead. As soon as he was old enough, he headed to London to seek training in mechanical drawing and architecture. He eventually designed buildings in New York City while earning additional income illustrating textbooks. Settling in New York in 1906, Paul was hired by fellow immigrant Hugo Gernsback (from Luxembourg) to illustrate his new magazine *The Electrical Experimenter* (later retitled *Science and Invention*). By the early 1920s Gernsback's publication began to feature science fiction stories accompanied by Paul's cover art and illustrations.

Science fiction suited Paul's talents, since he was free to envision buildings, rocket ships, and futuristic machines without the professional constraints of budgets and building clients. When Gernsback launched *Amazing Stories*, the first dedicated science fiction magazine, and promoted Paul to full-time illustrator in 1926, Paul became the first professional science fiction artist. As such, he was the first to paint a color image of a revolving space station (years before rocket scientist Werner von Braun proposed them), the first to depict plausible life on other planets, and to envision robot mineral explorers (which bear a surprising resemblance to today's Mars Rovers), two-way television communication (80 years before Skype) and other futuristic innovations.

Considering that Paul drew these at a time when most Americans still didn't even have telephones, it's hard to overstate the influence his work had on other thinkers of his day. His illustrations were the first science fiction images seen by Ray Bradbury, Arthur C. Clarke, Forrest J. Ackerman and others who would go on to great prominence in the field.

Paul died in Teaneck, New Jersey in 1963, and was posthumously inducted into the Science Fiction Hall of Fame in 2009. [Sources: Frank Wu website, and Frank R. Paul obituary, *New York Times*, 6/30/63]

....

City of the Future, ca. early 1940s, 13.5 x 30 in., airbrush and drafting inks, graphic arts paint. Courtesy of Dr. Stephen R. Turner.

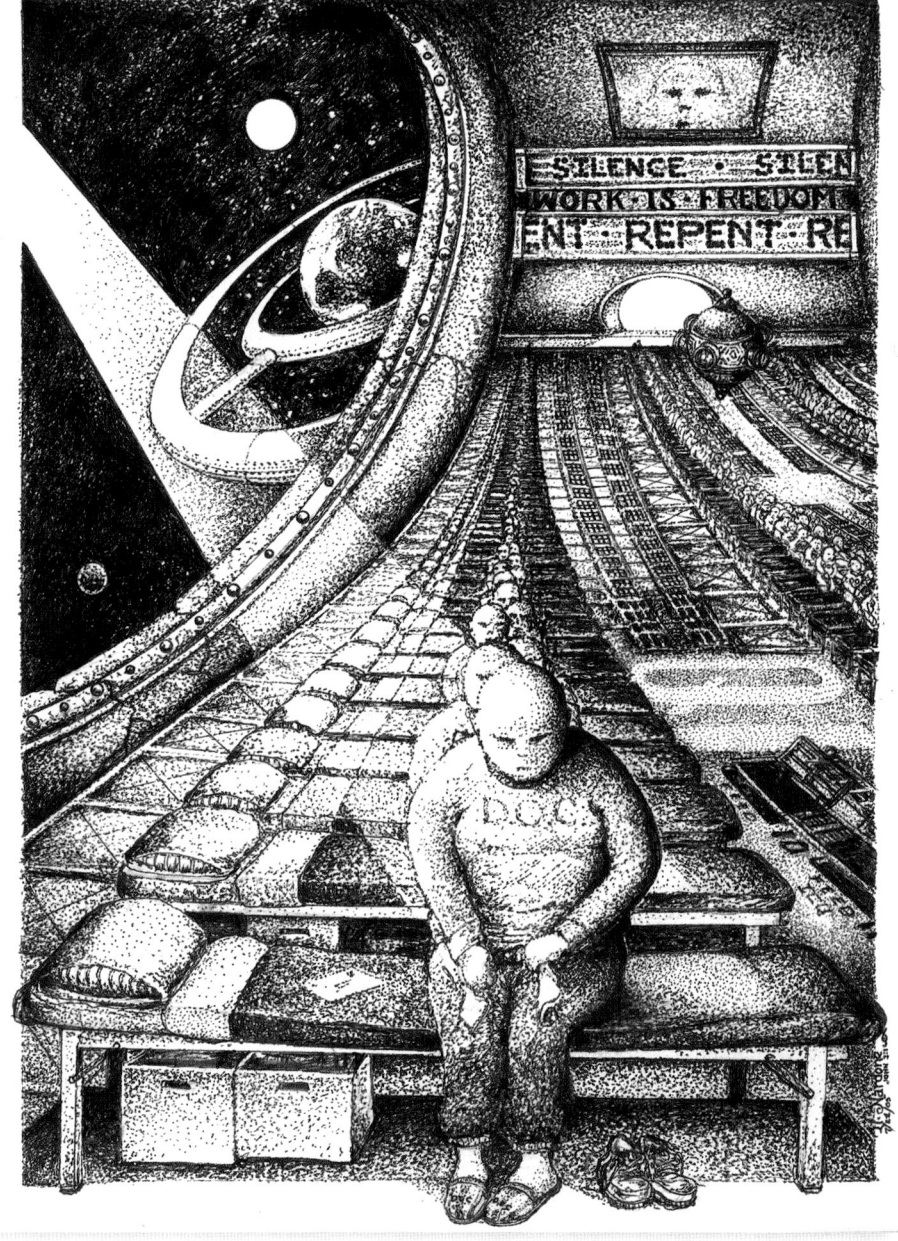

Vincent Nardone

Vincent Nardone was born in 1951 in Woburn, Massachusetts. At age 25 (in 1976) he was sentenced to life in prison for a crime committed in Maryland. During his first 25 years behind bars, Nardone made thousands of drawings on paper, strips of cloth, and on his cell walls focusing on just one narrow subject: clipper ships. When asked why he had chosen this particular theme, he would reply only that, "Everybody likes clipper ships."

In 1999, while participating in a Prison Arts Program at Osborn Correctional Institution in Somers, Connecticut, Nardone was prompted to try drawing a meaningful personal memory instead of a nautical scene. Inspired, Nardone is now renowned for the flood of extremely meticulous ballpoint pen drawings that flowed forth. These vivid and exactingly rendered drawings offered an escape from his circumstances. "When I'm doing my art," he says, "I don't exist in this place."

Nardone regularly contributes art to the Make-A-Wish Foundation, designs and makes toys for the Connecticut Department of Children and Family Services, and works as a mentor for the Re-Entry Program at Osborn. He hopes all his good works will eventually earn him parole. [Source: Jeffrey Greene, Community Partners in Action Prison Arts Program]

....

Exile (of the poor and beguiled), 2005, 11.75 x 8.75 in., ballpoint.
Courtesy of CPA Prison Arts Program.

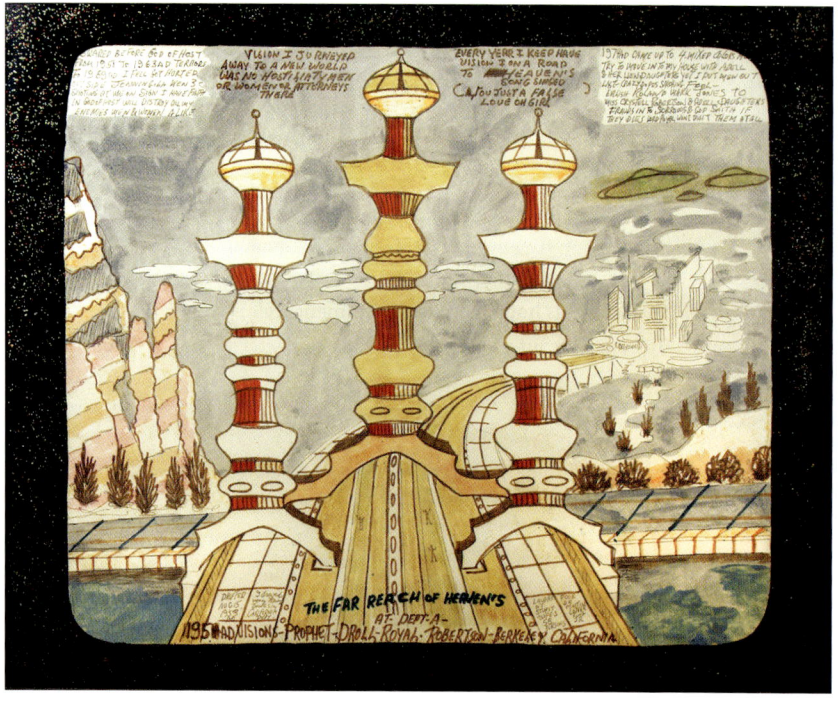

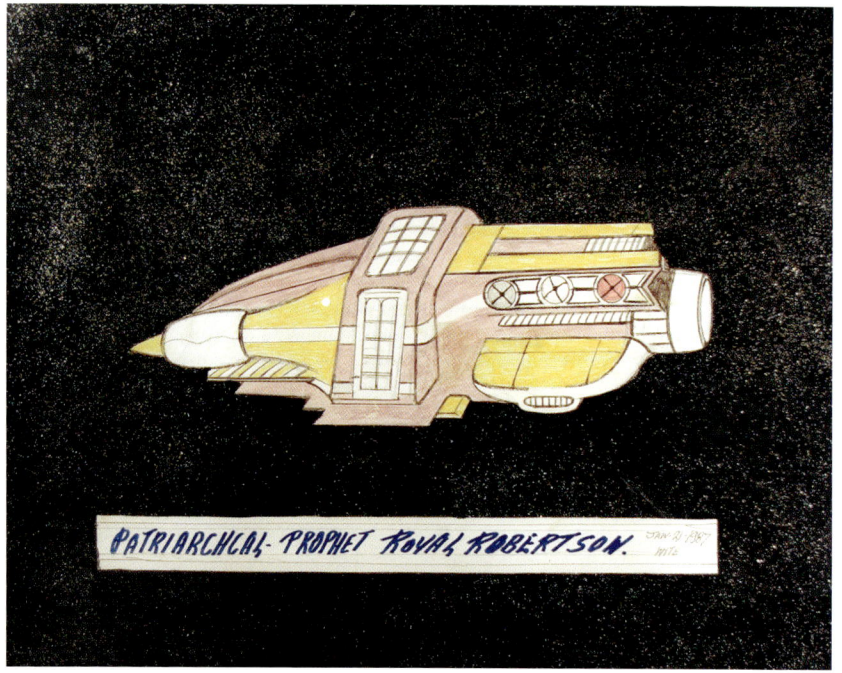

Prophet Royal Robertson

Royal Robertson was born in Baldwin, Louisiana, in 1936, and except for a few years apprenticing as a sign painter out west in his early 20s, he spent his entire life in Baldwin. He married Adell Brent in 1955. Despite difficulties, they managed to stay together for 19 years and raise eleven children. In the mid-1970s, however, the marriage fell apart and their children mostly sided with their mother. Not long after that, Robertson began lettering every available surface of his house with Old Testament style wrath.

He turned his entire front yard, along with all the exterior and interior surfaces of his house, into a vitriolic cry of blame, warning and protest. He called the environment "Artistico." Gradually expanding beyond Biblical references, he named himself the Libra Patriarch Prophet Lord Archbishop Apostle Visionary Mystic Saint Royal Robertson, and imagined himself the focus of an interplanetary female conspiracy.

Robertson began claiming to have gotten in touch with alien beings who supported his misogynistic campaign against adultery, fornication, lust and the women who made such sins possible. He created calendars and numerological charts that merged personal events from his marriage with predictions of the End Times, painted highly-detailed scenes of futuristic cities filled with Amazonian women (Adell among them) warring with space ships and flying saucers, and drew diagrams connecting deception, faithlessness, and treacherous female behaviors with patterns revealed by astrology and prophecy. In due time, he predicted, the Hurricanes of Hell would exact their revenge on such evildoers as his ex-wife.

Ironically, Hurricane Andrew flattened his home and all its signage in 1992. Robertson calmed and focused on rebuilding and recovery. He was in the process of reconciling with some of his estranged children when he died in 1997. [Sources: Anton Haardt, Robert Cargo, and Webb Gallery]

....

(opposite above) *The Far Reach of Heaven's*, ca. 1985,
21.5 x 26.5 in. Courtesy of Allen and Barry Huffman.
(below) *Rocket Car*, 1987, 21.5 x 27.5 in. Courtesy of Chris Roulhac
and William Sears. Both ink, watercolor, marker, pencil, glitter.

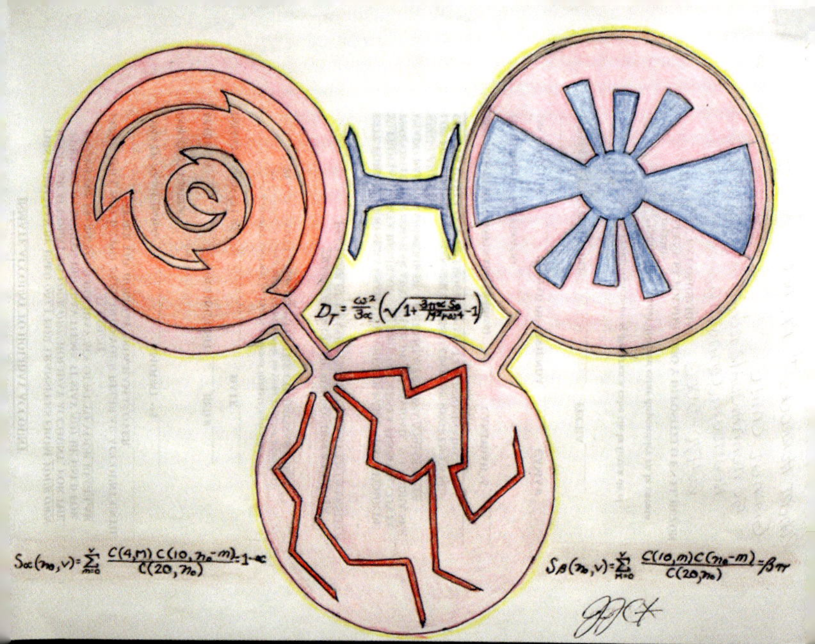

$$D_T = \frac{\omega^2}{3\kappa}\left(\sqrt{1+\frac{3\pi\kappa S_0}{H^2\tau_0 v}}-1\right)$$

$$S_\alpha(n_0,v) = \sum_{m=0}^{\infty} \frac{C(4,m)\,C(10,n_0-m)}{C(20,n_0)} - 1 = \infty$$

$$S_\beta(n_0,v) = \sum_{m=0}^{\infty} \frac{C(10,m)\,C(n_0-m)}{C(20,n_0)} = \beta\pi r$$

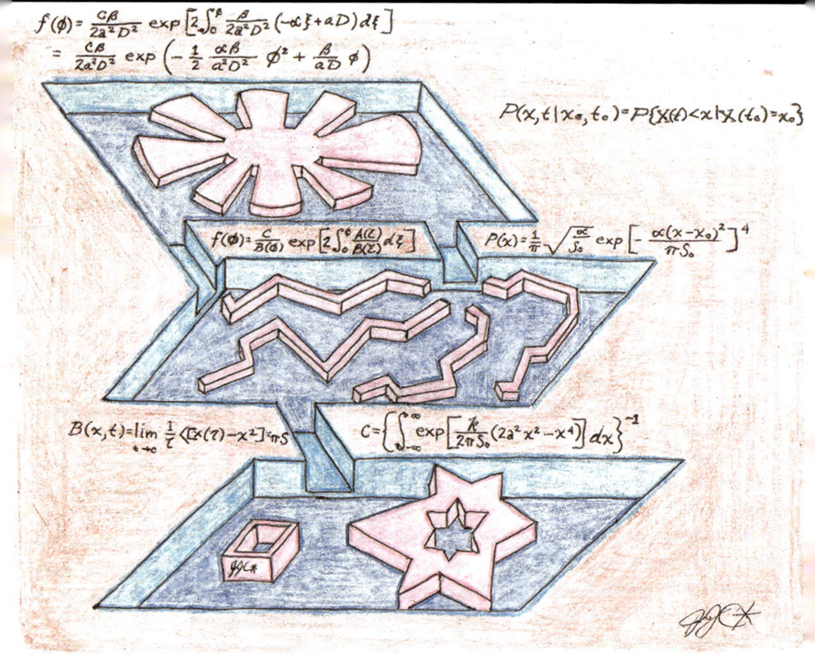

$$f(\phi) = \frac{C\beta}{2a^2 D^2}\, \exp\left[2\int_0^\phi \frac{\beta}{2a^2 D^2}(-\alpha\xi + aD)\,d\xi\right]$$
$$= \frac{C\beta}{2a^2 D^2}\, \exp\left(-\frac{1}{2}\frac{\alpha\beta}{a^2 D^2}\phi^2 + \frac{\beta}{aD}\phi\right)$$

$$P(x,t\,|\,x_0,t_0) = P\{X(t) < x \,|\, X(t_0) = x_0\}$$

$$f(\phi) = \frac{C}{B(\phi)}\, \exp\left[2\int_0^\phi \frac{A(\zeta)}{B(\zeta)}\,d\zeta\right]$$

$$P(x) = \frac{1}{\pi}\sqrt{\frac{\alpha}{S_0}}\, \exp\left[-\frac{\alpha(x-x_0)^2}{\pi S_0}\right]^4$$

$$B(x,t) = \lim_{t\to\infty} \frac{1}{t}\langle[x(\tau)-x^2]^2\rangle\,\pi S$$

$$C = \left\{\int_{-\infty}^{\infty} \exp\left[\frac{\hbar}{2\pi S_0}(2a^2 x^2 - x^4)\right]dx\right\}^{-1}$$

Joseph Castellano

Joseph Castellano (b. 1949) is a Connecticut prison inmate who began serving a 60-year sentence for felony murder in 1993. He uses pens and color pencils to create drawings that reference chemistry, physics and magic in an effort to convey his ideas about the universe's inner workings.

Castellano's art began to reach an audience beyond the prison walls after 2005, when he submitted works to the annual exhibition of the Prison Arts Program sponsored by Community Partners in Action (CPA), a non-profit agency headquartered in Hartford, Connecticut, with a mission to assist people affected by the criminal justice system. Castellano has become one of the most artistically prolific inmates in the program.

His drawings titled *Universal Time Clock* and *When Then Was Now* represent attempts to explain the nature and function of time, a common and constant fixation for many inmates. Scientific equations are central to his work, which he often accompanies with poetry. [Source: Jeffrey Greene, Community Partners in Action Prison Arts Program]

(opposite above), *Universal Time Clock*, 2005,
8.5 x 11 in., ink and color pencil on inmate account form.
Courtesy of CPA Prison Arts Program.

(below) *When Then Was Now*, 2005, 8.5 x 11 in,
ink and color pencil. Courtesy of Jeffrey Greene.

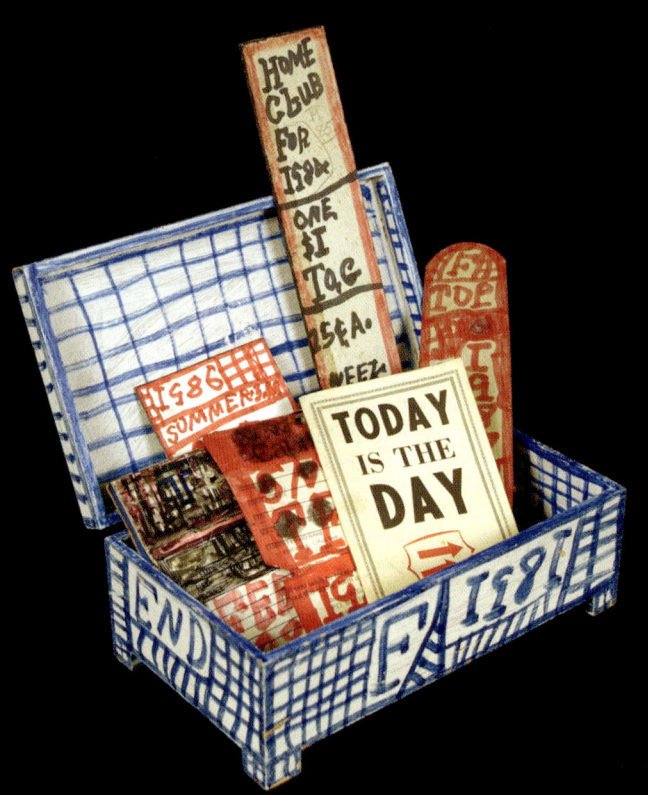

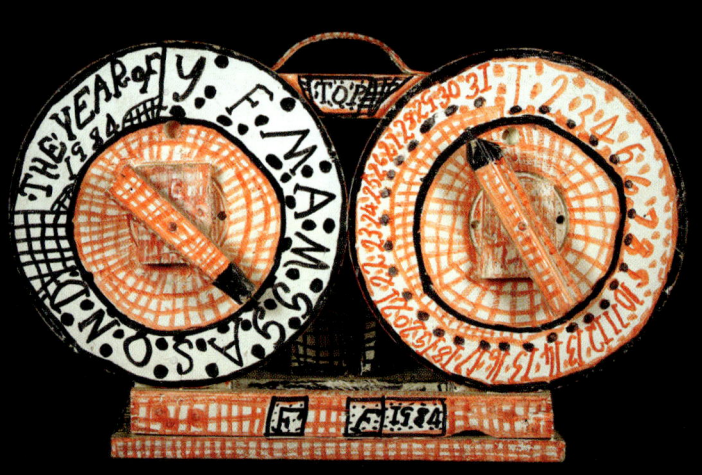

Zebedee B. Armstrong

"Z.B." Armstrong was born in Thomson, Georgia in 1911. He attended school through the 8th grade, married in 1929 and raised two daughters. For most of his adult life, Armstrong picked cotton on the same farm where his father had worked, but following the death of his wife in 1969 until his retirement in 1982, Armstrong worked at the Thomson Box Factory.

In 1972 he had a vision about the end of the world. After that, he began building many box-like computational devices and calendars with which he hoped to be able to calculate the exact date of the End and Final Judgment.

He realized that he could profit from pinpointing the date, and developed a secret speculative scheme that would pay out on the day *after* the world was predicted to end. Dozens of his neighbors invested a few dollars a week, unaware that Armstrong intended to have the last laugh by "beating the clock." As far as anyone knows, he never paid anyone back. His own world ended in 1993, the year he died. [Sources: Louanne Laroche, Tom Wells and author interviews with the artist, 1987]

Components of the *Future Predictor*, 1970–85, (opposite above),
9 x 9.25 x 8 in.; (below) 15 x 23.5 x 6 in.; both paint and marker on
cardboard, wood, and found objects. Courtesy of the LaRoche Collection.
(below) Roger Manley, photo of Armstrong at home in Thomson, GA, ca. 1985.

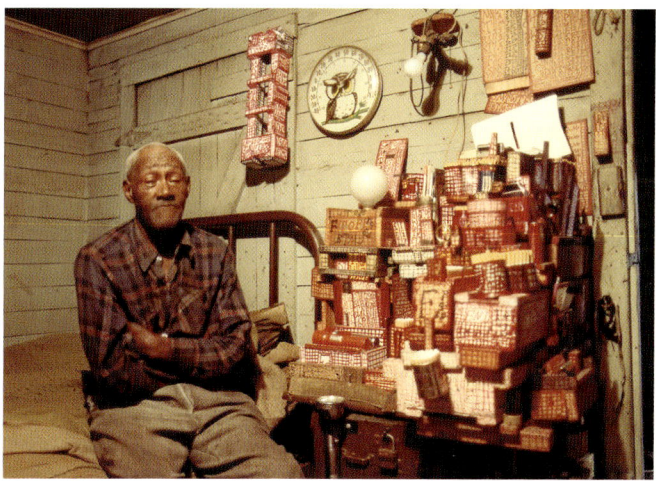

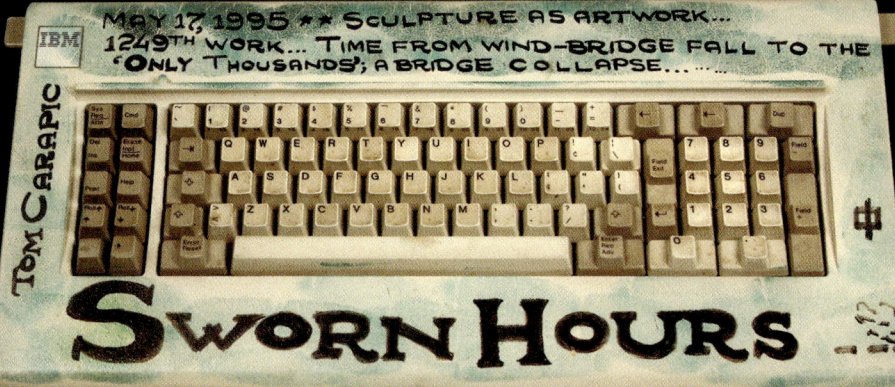

MAY 17, 1995 ** SCULPTURE AS ARTWORK...
1249TH WORK... TIME FROM WIND-BRIDGE FALL TO THE
"ONLY THOUSANDS"; A BRIDGE COLLAPSE..."...

TOM CARAPIC

SWORN HOURS

Tom Carapic

Tomislav Sava Carapic not only covered all the walls and furniture of his small New York apartment with his artwork on newsprint (signed with his thumb or handprints) but swathed much of the surrounding midtown neighborhood as well.

Born in 1939 in the village of Velesevec, Croatia (then Yugoslavia), he was sent to a military school in preparation for becoming a sergeant in the Yugoslavian People's Army. College was denied to him because he wasn't a Communist Party member, so he crossed the Italian border illegally in 1961 and managed to get to America a year later.

In 1965, Carapic began taking art lessons at New York's Art Students League but soon dropped out because the classes seemed boring. They were "like getting a vacation for 10 million years," he said. He enrolled in hairdressing and cosmetics classes at the Wilfred Academy of Beauty Culture instead, but found so little work as a stylist that he barely made ends meet as a slitting-machine operator at the Noesting Pin Ticket Company.

In the late 1970s, Carapic began suffering stress-induced hallucinations after he had trouble earning enough credits for a degree at Manhattan Community College. He believed that the college had been invaded by troops from "the evil marriage bureau" who had conducted "an Air Force bombardment" of the school. Soon after these visions, he began making art from computer components and other found electronics parts, especially IBM keyboards. His most famous exhibit in New York City was "Big Bang Theory," which included doomsday warnings painted on computer keyboards, shoes and construction debris. [Source: Gareth Brown, Outsider Gallery]

....

Sworn Hours, 1995, 1.75 x 19 x 9 in., marker, paint, ink on computer keyboard.
Courtesy of American Primitive Gallery, New York

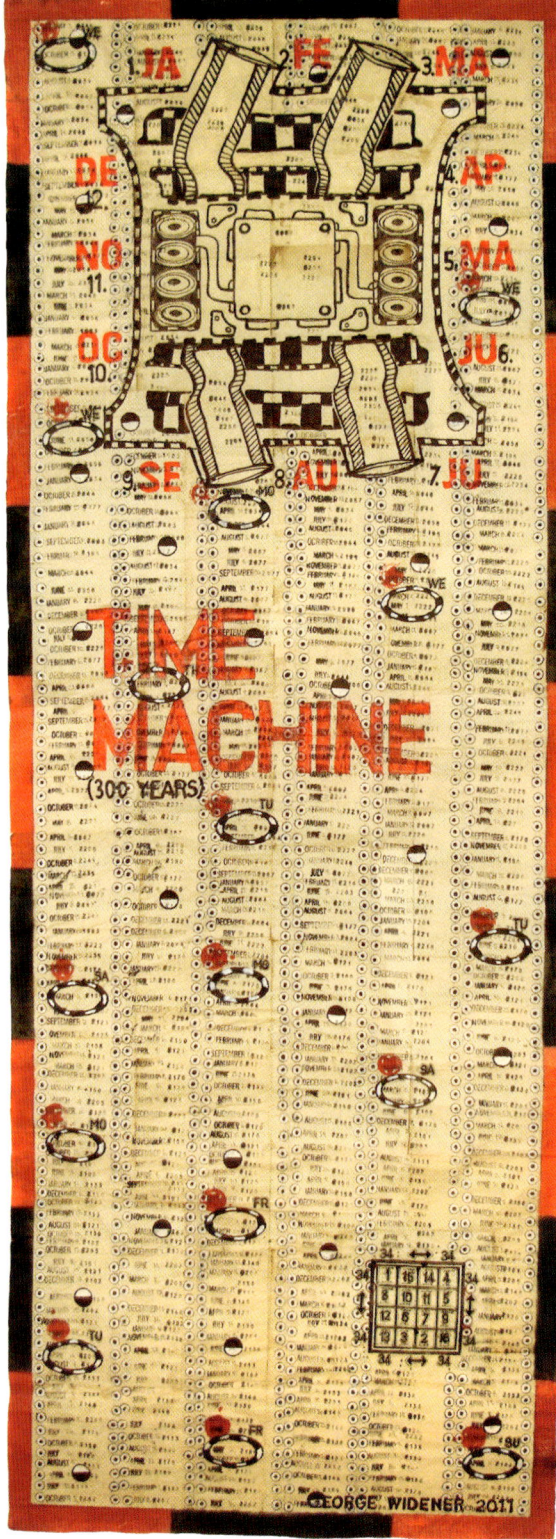

George Widener

George Widener (b. 1962) lived with his parents and five siblings in Cincinnati, Ohio until he was 10, when his father died of a heart attack. Sent to live with relatives in Kentucky and Tennessee, he was an awkward and sometimes violent teen, prone to drawing or exercising his exceptional abilities at mathematical calculation, memorization and virtually anything to do with numbers and calendars.

After graduating high school with honors at 17, he served four years in the Air Force as an intelligence technician, specializing in aerial photo-surveillance. Discharged in 1984, he briefly studied engineering at the University of Texas, but dropped out due to personal and financial troubles that also left him homeless for the next ten years. Working odd jobs to survive, Widener wandered the streets of Amsterdam, London, Berlin and Paris filling notebooks with drawings of urban buildings, numbers, calendrical sequences, facts and statistics.

In the early 1990s he withdrew further and experienced intense anxiety about his drawings and notations. After being diagnosed with a high-functioning form of Asperger's Syndrome, he was able to earn a liberal arts degree from the University of Tennessee in 1998. His drawings became larger and more heavily worked, combining architectural renderings with lists of dates, numbers, and statistical information.

Widener's concerns include a more humane approach to urban design (exemplified in his "Megalopolis" series) and what he calls the "Singularity"—the moment that artificial intelligence will surpass human intelligence, enabling machines to take over the Earth. He incorporates "magic time squares"—numerical grids in which rows, columns, and diagonals of calendar dates add up to identical sums. Widener's most detailed drawings feature dense fields of numerical dates which he claims can only be appreciated by super-computers that haven't been built yet. [Sources: The artist, Henry Boxer Gallery, and Roger Cardinal, "The Calendars of George Widener," in *Raw Vision* #51]

....

Time Machine, 2011, 63 x 23.5 in., rubber stamp ink, watercolor
on paper towels. Courtesy of Henry Boxer Gallery, London.

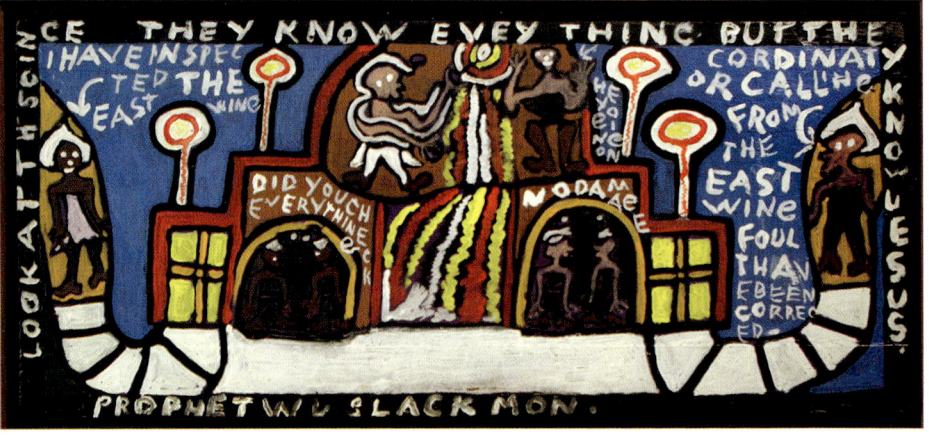

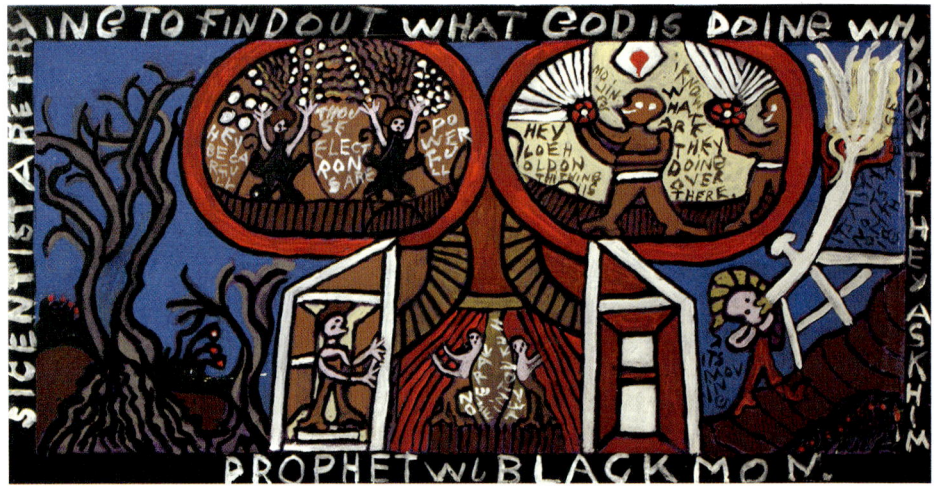

William Blackmon

In 1921, William Joshua Blackmon was born a seventh son, a birth order often associated in traditional folklore with an ability to prophesy. He grew up on a small farm near Albion, Michigan. Even as a boy he was credited with that ability, foretelling the death of a neighbor and other events. In 1937 he dropped out of high school and went to work, first on a track gang for the New York Central Railroad, and then for the Albion Malleable Iron Works, a hellish foundry.

World War II took Blackmon to New Guinea and the Philippines where he fought with the Army's 585th Engineering Company and earned six bronze stars. He "found religion" in the foxholes, where only intense prayer seemed able to overcome his fears of enemy air attacks.

After the war, Blackmon ran a shoeshine parlor and worked as a hitch-hiking sidewalk street preacher in Chicago. His first wife died in childbirth and a second wife abandoned him. In 1974, after cutting open a watermelon and seeing the letters "M I" spelled out in the pattern of seeds, he took it as a message from God and moved to Milwaukee, where he started a laundry and sewing shop he called the Revival Center Shoe Repair and Shine Parlor. Later he added a storefront church and self-help agency with a rummage sale component, renaming it the "World Revival & Interprice Center."

After his hand-lettered signage began to attract local attention, Blackmon started illustrating subjects from his street sermons by painting on scrap lumber. His central theme was a warning against the destructive path of secular technology such as hospital medicine, science, television, sex education, nuclear power and other features of modern life. "You see them on TV," he would say, with "all their great inventions and things, and with all they do, you know you will never hear one word: God! That's right! He is the one who should be getting the credit, because He's the one who made those things possible." In 2010, Prophet Blackmon died in the Veterans Administration hospital in Detroit and was buried nearby with military honors.

[Source: Jeffrey R. Hayes, et al. "Prophet William J. Blackmon: God's Modern-Day (Artist) Apostle" in *Signs of Inspiration: The Art of Prophet William J. Blackmon*. Haggerty Museum of Art, Marquette University, Milwaukee]

....

(opposite above) *Look at the Scince They Know Everything But . . .* ,
ca. 1985, 13.5 x 30.75 x 1 in. Courtesy of John Foster.
(below) *Sicentist are trying to find out what God is Doing*, ca. 1985,
13.5 x 26.5 x 1 in. Anonymous lender. Both are oil enamel on wood.

CHECKLIST & INDEX

All measurements are in inches, and all 2D works are on paper unless noted otherwise.

Charles Carroll, 64–65

City, ca. 1986, 2.25 x 2.5 x 2.75, balsa wood, acrylic, beveled glass box; *City*, ca. 1990, 4.75 x 4 x 4, balsa wood, acrylic, marbleized paper, wood, glass dome; *City VII*, ca. 1990, 4 x 18 x 18.5; balsa wood, acrylic; *City X*, 1997–98, 4.5 x 21 x 18.5; balsa wood, acrylic; *City XXVI*, 1997–98, 13.75 x 10.5 x 8, balsa wood, acrylic [collection of the artist]

Joseph Castellano, 104–105

Universal Time Clock, 2005, 8.5 x 11, ink and color pencil on inmate account form [CPA Prison Arts Program]; *When Then Was Now*, 2005, 8.5 x 11, ink and color pencil [Jeffrey Greene]

Silvain Corentin, 60–61

Rome Tower, 2011, 90 x 13 x 17, found wood, ceramics, paint [Cavin-Morris Gallery]

Charles Dellschau, 46–47, back cover

Airship 44–74, 1919, 19.25 x 15.5; *Airship 46–77*, 1920, 17.25 x 16.75; *Airship 45–89*, 1920, 17.75 x 16.25; *Airship 4753–75*, 1920, 13.25 x 17, [Stephen Romano]; *Airship 44–99*, 1919, 16.75 x 17 [Selig D. Sacks]; all watercolor, poster paint, pencil, collaged news clippings

John Devlin, 66–67

New Cantabrigiensis drawings, 1988, all 8.5 x 11, color pencil, crayon, markers, collaged photos [Henry Boxer Gallery]

Jacques de du-Glass, 62–63

Street plan of Lynxbourgh, n.d., 12 x 18, ink and color pencil [Lindsay Gallery]

Energo S.p.A, Torino, front cover, iv

Shock Helmet, ca. 1900, 23 x 12 x 13.5, wood, metal, rubber, copper coil [Steve Erenberg, radio-guy.net]

Paul Esparza, 97

Metro Link Farebox & More, 1995–96, 19.5 x 15.5; *Future City*, 1995–96, 19.75 x 15.5; *Trains & More*, 1995–96, 9.5 x 12.75; all acrylic on canvas panels [Pardee Collection]

Howard Finster, 5–6, 20, **38–39**, 81, 91

The Whole World Passing Between the Two Superpowers in Fear, 1985, 21.5 x 26.75, monotone lithograph. [anonymous lender]; *Heavenly Mansions*, n.d., 16 x 11.75, color marker and pencil; *Find Out Knowledge of Witty Inventions*, n.d. (Finster #1094), 17.5 x 31, enamel on panel, pyrographic frame; *Flying Saucer Car*, n.d., 12 x 9.25, ballpoint on graph paper; *The One Wheel Course Power Drive*, n.d., 12 x 9.25, ballpoint on graph paper; *All Things Speak for God and His Word,* ca. 1960s, 9 x 6, ballpoint on butcher paper; drawing for invention, 5.75 x 8, graphite and ballpoint [Jane and Bert Hunecke]; *Vision of Moddle Structures*, 1987, Finster # 6617), 23.75 x 47.25 in., enamel on panel [John Denton]

Stephen Gessig, 42–43

Inventor's box, n.d., 9 x 15 x 3.5, wood, glass, screen, crayon and ink on cardboard [American Primitive Gallery]

Henry Hill, 74–75

Molecular Concourse / Rampart and Rafter, ca. 1960, 20 x 32, graphite pencil [anonymous lender]

Chris Hipkiss, 84–85

Anchusaazurea / Our Slit with the South, 2004, 36 x 58.5; *Wolfe & Stole, Inc.*, n.d., 33.25 x 23.5, both graphite pencil [Cavin-Morris Gallery]

Renaldo Kuhler, 58–59
Monuments to Casar Nicolai and Capt. Zumbakar, ca. 1960, 8.5 x 11, ink; *Roca-terranian National Museum,* ca. 1960–65, 11 x 14, color pencil, watercolor; *Untitled* (farm with airship), ca. 1960, 11 x 14, watercolor; *Untitled* (planetarium projector), ca. 1960, 11 x 8.5, ink and color pencil; *Casa Victoria,* ca. 1955, 11 x 8, India ink [Brett Ingram]

Paul Laffoley, 1–2, 26, **82–83**
Das Urpflanze Haus II (1–4), 1993–95, 27 x 36, drafting ink and presstype; *Das Urpflanze Haus II (5–8),* 1993–95, 27 x 36, drafting ink and presstype; *Die Erde Blume,* 1994, 27 x 19.5, wood, mylar, glass, drafting ink, photo collage, pres-stype, watercolor [Cartin Collection]; *Das Urpflanze Haus* model, 1997, 25.5 x 44.5 x 63.75, mixed media [collection of the artist]

Duncan Laurie, 10–14, **30–33**
Purr Generator (with Gordon Salisbury and Todd Thille), 2002 and 2011, 85 x 118 x 118, mixed media, aluminum, elec-tronic circuitry, LED display; *Bio-sen-sors,* 2001, 20.5 x 13, glass, felt, wire mesh, wires, fungi, stones; *Radionic Cir-cuit with Sheep,* 1993, 13 x 9.5; *Radionic Hand,* 1993, 7 x 6; *Hieronymus Circuit,* 1994, 5.5 x 6.5; *Electronic Circuit With Crop Circles,* 1993, 7 x 6; *Multi-Wave Oscillators,* 1994, 17 x 16; all are printed circuits [collection of the artist]

Alex Maldonado, 24–25, **50–51**
Giant Vacuum (Tower) 800-ft. High, 1987, 26 x 32; *Instant Justice,* 1980, 19.25 x 23.5; *Underground Caves,* 1971, 20 x 24; *Maldonado's Giant Auditorium,* n.d., 23.25 x 27.5; *Memorial To Martin Luther King, Jr.,* 1987, 27.75 x 33.75; *21st Century—5 Telescopes In One—X-ray Computer,* 1986, 21 x 27, all oil on canvas or canvas board with painted frames [Ames Gallery]

Eddie Owens Martin (Saint EOM), 4–6, **86–87**
Pasaquoyan Man in Space in the Future, ca. 1950, 29.5 x 21.5, oil and metallic enamel on panel; *Untitled* (man comb-ing hair), ca. 1960, 21.5 x 29.5, poster paint on panel [Pasaquan Preservation Society, Inc.]

Vincent Nardone, 100–101
Exile (of the poor and beguiled), 2005, 11.75 x 8.75, ballpoint [CPA Prison Arts Program]

Melvin Edward Nelson, 16, **48–49**
Untitled Futurist Visions, ca. 1965, graphite pencil, poster tempera, water-color [Cavin-Morris Gallery]

Sean Pace, 2–3, **56–57**
Death Slapper, 2008, 41 x 64 x 26.5, post drill, electric motor, rubber chick-ens, plastic skull, mixed media; *Social Landscape Painting Machine,* 2011, 65.5 x 54 x 112, bicycle, road signs, file box, mixed media [collection of the artist]

Frank R. Paul, 98–99
City of the Future, ca. 1940–45, 13.5 x 30, airbrush and drafting inks, graphic arts paint [Dr. Stephen R. Turner]

Achilles Rizzoli, ii, **68–69**
Mother into Stone Proemshayed, 1939,
58 x 34, graphite pencil, purple ink;
Mother in Metamorphosis Idolized,
1938, 54 x 35, graphite pencil, drafting
ink; *Virginia Ann Entwistle Symbolically
Sketched*, 1944, 35.5 x 23.5, cyanotype
blueprint; *Y.T.T.E. Plot Plan—Fourth
Preliminary Study*, 1938, 39 x 25, ink,
watercolor, color pencil [Ames Gallery]

Royal Robertson, **102–103**
The Far Reach of Heaven's, ca. 1985,
21.5 x 26.5, [Allen and Barry Huffman];
Rocket Car 1987, 21.5 x 27.5, [Chris
Roulhac and William Sears]; both ink,
watercolor, marker, pencil, glitter

André Robillard, **54–55**
Fusil x USA A.W.R. Rapide T136, ca. 1985,
22 x 45 x 8 in., wood, found objects,
mixed media [anonymous lender]

Sabato Rodia, **76–80**
Photograph of the *Watts Towers*, 1987,
27.75 x 39.25, inkjet print, by Ted
Degener; scale model of the *Watts Towers*, 1997, 27.5 x 49 x 19.5, painted and
glued matboard, wood, by Larry Harris
[collections of the artists]

Welmon Sharlhorne, **72–73**
Untitled (building), n.d., 19 x 22.25,
color ink on gray paper; *Untitled*
("surreal building"), n.d., 8.5 x 14, ink
[American Primitive Gallery]

Lee C. Spooner, 19, **44–45**
Patent Drawings for self-propelled/per-
petual-motion applications, 1911–1935,
up to 15 x 19, graphite and ink on
found paper and fabrics catalogue [The
Museum of Everything]

Ionel Talpazan, 7–8, **88–89**
UFO Tearis Fundamental, 1997, 29 x
65, poster paint, pencil, ink, oil crayon
[American Primitive Gallery]; *UFO* (with
silver/white outlines), 1999, 30 x 63,
poster paint, metallic paint, pencil, ink,
oil crayon and *UFO* (with blue thruster
jet at right), 30 x 57.75, poster paint,
pencil, ink, oil crayon [Jack Stevens];
UFO (with two spiral emanations at
right), 1999, 30 x 61, poster paint,
pencil, ink, oil crayon on paper [Henry
Boxer Gallery]

Melvin Way, 18, **28–29**
F'CPE, n.d., 5 x 4, red ink, marker; *Loki*,
n.d., 2.75 x 2.75, black ink; *Kama Ere
God of Love*, 2009, 11.5 x 8.5, color pencil and ball point; *The Schwahn*, n.d. 2 x
5, red and black ink; *Iron Man 2*, n.d., 4
x 5.5, black ink; *Sun Star*, n.d., 4 x 5.5,
black ink; *Tantric Sect*, n.d., 5.75 x 5.25,
black ink, collage; *ME=Mat/2*, n.d., 4.75
x 8.75, black ink; *Cogito Ergo Sum*, n.d.,
5 x 3, black ink; *Mustaphau*, n.d., 5.5 x
3.75, dark blue ink [Gallery at HAI]

George Widener, 8–9, **110–111**
Time Machine, 2011, 63 x 23.5, rubber
stamp ink, watercolor on paper towels
[Henry Boxer Gallery]

James Wong, **92–93**
Dead War Toys, 2011, 18.5 x 23.5;
Upgrade Hanger 79B, n.d., 18.5 x 23.5;
Untitled, 2009, 17.5 x 23.5, all ink and
marker on Bristol board [American
Primitive Gallery]

STAFF

Roger Manley, *Director*

Zoe Starling, *Curator of Education*

Mary Hauser, *Registrar*

Teppei Umeno, *Assistant Registrar*

Matt Gay, *Art Preparator*

Hilary Kinlaw, *Museum Operations Manager*

Clara Ray and Rebekah Velasquez, *Security and Reception*

Janine LeBlanc, *Textiles Consultant*

Photography: Roger Manley

Catalog design: Amy Ruth Buchanan, 3rd sister design

Editor: Chris Vitiello

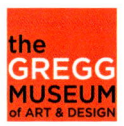

GREGG MUSEUM OF ART & DESIGN

NCSU Campus Box 7306 | Raleigh, NC 27695-7306

www.ncsu.edu/Gregg | 919.515.3503

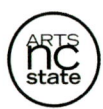 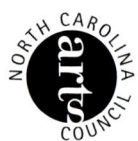 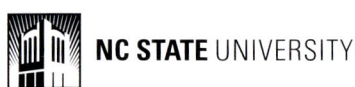